Biotechnology and the Patent System
Balancing Innovation and Property Rights

Claude Barfield and
John E. Calfee

The AEI Press

Publisher for the American Enterprise Institute
WASHINGTON, D.C.

Distributed to the Trade by National Book Network, 15200 NBN Way, Blue Ridge Summit, PA 17214. To order call toll free 1-800-462-6420 or 1-717-794-3800. For all other inquiries please contact the AEI Press, 1150 Seventeenth Street, N.W., Washington, D.C. 20036 or call 1-800-862-5801.

Library of Congress Cataloging-in-Publication Data

Barfield, Claude E.
  Biotechnology and the patent system : balancing innovation and property rights / Claude Barfield and John E. Calfee.
      p. cm.
    ISBN-13: 978-0-8447-4256-4 (pbk.)
    ISBN-10: 0-8447-4256-2
    1. Biotechnology—United States—Patents. 2. Patent laws and legislation—United States. I. Calfee, John E., 1941- II. Title.

      KF3133.B56B37 2007
      346.7304'86—dc22

                                                              2007035150

11 10 09 08 07   1 2 3 4 5

Printed in the United States of America

Biotechnology and the Patent System

Contents

Acknowledgments

We gratefully acknowledge the support of the Biotechnology Industry Organization (BIO) for early research on this project, though the judgments rendered in the study are those of the authors alone. We would like to thank the following people for counseling us or commenting on parts or all of this study: Bruce Lehman, Kenneth Dam, John Thomas, Keith Maskus, Josh Lerner, Hayden Gregory, Theodore Frank, Michael Greve, Steven Merrill, and David Mowery. In addition, Andrei Zlate made major contributions to an earlier version of the study. We are indebted to Zach Rosenthal and Dan Geary for fact-checking and other administrative tasks. Any remaining errors in judgment or fact are, of course, ours.

Introduction

On April 29, 2007, a unanimous Supreme Court handed down what the *New York Times* called "the most important patent ruling in years," mandating stricter standards for obtaining patents that combine elements of preexisting inventions (Greenhouse 2007). Two weeks before that decision, the chairmen and ranking minority members of the Senate and House Judiciary Committees introduced legislation for sweeping changes in the patenting process in the United States. These two events will immediately and in the more distant future have momentous consequences for the U.S. biotechnology industry and allied R&D community. This study evaluates proposals for overall patent reform through administrative and legislative routes, and advances our own recommendations for meaningful but prudent change, with the goal of achieving an efficient and equitable balance between the rights of patent-holders and those of challengers.

Biotechnology—the incorporation of biological mechanisms into technology—is as old as plant and animal breeding, dating back thousands of years. The modern biotechnology industry, based on startling advances in molecular biology, is but a few decades old, however, and still in its childhood. Already it has created extraordinary treatments and diagnostics, ranging from HIV tests to targeted cancer therapies that are revolutionizing oncology. Even so, the industry as a whole remains unprofitable as it undertakes massive investments with highly uncertain returns. Like the larger pharmaceutical industry of which it is a part, the biotechnology industry relies upon intellectual property protections, primarily patents. Most biotechnology inventions require large investments to discover which products can meet U.S. Federal Drug Administration (FDA) approval requirements and then be used by patients. Without patent protection, investors would

1

see little prospect of profits sufficient to recoup their investments and off-set the accompanying financial risk.

Biotechnology patents, especially the so-called gene patents, pose difficult economic and legal issues. In addition to simply permitting temporary monopoly pricing of new inventions, patents can support efficient exploita-tion of a research stream; but a multiplicity of related patents can impose substantial transaction costs that impede progress. The patent system and its users have exhibited powerful self-correcting forces, however, which to date have largely preserved the inherent benefits of biotechnology patents while imposing, at most, modest inefficiencies. As they move forward with patent legislation in the 110th Congress, our first admonition to congres-sional leaders is, "First, do no harm," as the recent history of patent system "reform" is replete with examples of unexpected—and unintended—negative consequences.

That said, we will argue that the most important changes Congress can enact are those that allow more information and expertise to be gleaned and utilized by the U.S. Patent Office and patent examiners as they render their decisions. Thus, the core of our recommendations aim to introduce "bounded adversarial" elements into the patent application process and the post-grant appeals process—"adversarial," to utilize the expertise and arguments of outside opponents to alert the patent office to prior art or defi-ciencies in the patent application; and "bounded," by tight procedural rules, to allow all sides to present their cases while protecting against frivolous challenges whose main goal is to gum up the system and ultimately wear down (and "hold up") the patent-holder.

1

Biotechnology and Health

A remarkable confluence of purely intellectual advances and new technologies gave rise to, first, an enticing prospect of revolutionary advances in medical practice and, second, a small, highly entrepreneurial industry bent upon devising medical products that could pass the U.S. Food and Drug Administration's rigorous drug approval process and gain the confidence of practicing physicians. The results have begun to transform medical practice.

The Emergence of the Modern Biotechnology Industry

During the 1970s and '80s, a new industry began to exploit biological processes rather than traditional chemical methods to develop drugs, vaccines, and diagnostic tests. Biotechnology itself—the harnessing of biological mechanisms to technology—was hardly new. Plant and animal breeding, for example, had been used for thousands of years and had achieved such remarkable successes as the "Green Revolution" in agriculture (Pimentel 2004). But the "new biotechnology" directly addressed cellular and biomolecular processes to solve problems and invent new products (Biotechnology Industry Organization 2005, 1) that began to transform the pharmaceutical industry and, thus, health care itself.[1] At the same time, the discovery and isolation of genes that govern the creation of therapeutic (or hostile) proteins and the development of such tools as genomic arrays also began to revolutionize traditional drug development, further extending the reach of science toward a multitude of previously untreatable diseases and conditions.

The first biotechnology drug—human insulin produced in genetically modified bacteria—was approved in 1982. Since then, the biotechnology

3

industry has created more than four hundred approved drugs and vaccines for cancer, Alzheimer's disease, Parkinson's disease, rheumatoid arthritis, heart disease, diabetes, multiple sclerosis, AIDS, and psoriasis, among others (Biotechnology Industry Organization n.d.) By 2003, more than 15 percent of the top-selling two hundred drugs worldwide had been developed by biotechnology companies, alone or in partnership with the pharmaceutical industry (Ernst and Young 2003). As of 2005, more than a hundred biotech products involved recombinant DNA or monoclonal antibodies, tools that did not exist before the mid-1970s. Worldwide sales of recombinant DNA drugs were $40 billion in 2005 (InfoService Biotechnology 2007). Most biotech drugs are relatively new; 210 were approved by 1999, and another 210 were approved between 2000 and 2005 (Biotechnology Industry Organization 2007b, 4).

Biotechnology has also produced hundreds of new diagnostic tests, many of them based on the identification of genetic mutations in human DNA. DNA sequences can now be used to identify individuals predisposed to particular diseases and to detect conditions early enough to be successfully treated. Prominent among biotechnology-based diagnostics are tools to identify genetic susceptibility to breast or ovarian cancer, to detect pathogens such as streptococcus or HIV in the blood supply, and to support home pregnancy tests (Biotechnology Industry Organization 2005, 41). Additional progress has been made on diagnostics for chlamydia and gonorrhea, congestive heart failure, hepatitis B and C, and hundreds of other targets.

In their annual review of the biotechnology industry, the consulting firm of Ernst and Young (2007) revealed an American industry consisting of 1,456 firms, of which 336 were publicly traded, with a total U.S. market capitalization of about $400 billion. U.S. revenues grew from $8 billion in 1992 to $51 billion in 2005 (Biotechnology Industry Organization 2007b, 2). Last year, publicly traded firms in the United States alone employed 131,000 people (Ernst and Young 2007, 7). Immense amounts of private financing—nearly $100 billion in the years 2000–5 (Biotechnology Industry Organization 2007b, 5) and $20.3 billion in 2006 (Ernst and Young 2007, 7)—have been invested in this industry. Nonetheless, the biotechnology industry as a whole incurred net losses of $5 billion in 2006. This was largely because of the extraordinary costs of developing new drugs. The biotech pharmaceutical industry spent about $28 billion on research in 2006 (Ernst and Young 2007, 7).

The First Wave of Modern Biotechnology

Biotechnology-based advances in medical therapy have simplified medical procedures, reduced health-care costs, increased workplace productivity, alleviated pain and suffering, improved the lives of the elderly, and increased life spans. An example is Gleevec (imatinib mesylate), approved as one in the new generation of targeted cancer drugs. It treats chronic myeloid leukemia and costs about the same as traditional chemotherapy (which involves substantial clinical and hospital costs for treatment and monitoring), but its side effects are far less debilitating, often allowing patients to lead relatively normal lives, including being employed.[2] Like most biotech-based cancer drugs, Gleevec was quickly tested on other cancers.

A substantial body of economic research has demonstrated the broad benefits of new pharmaceuticals, including many biotechnology products. A Florida Medicaid study of AIDS patients, for example, found that after a special waiver permitted the use of newer, more expensive drugs, annual pharmaceutical expenditures increased by about $560 per patient, but overall costs decreased by more than $800 (Kleinke 2001, 47). Another Medicaid study, this one on the effects of limiting patient access to schizophrenia drugs, found that reduced usage of newer treatments increased total health-care costs (Soumerai et al. 1994). Other research has examined broad trends in such relationships. Taking numerous potential confounding factors into account, Lichtenberg (1996, 2002a, 2002b, 2003) found substantial savings from health-care improvements associated with the adoption of new drugs. Far more important may be the savings to society as a whole. A recent evaluation of routine child vaccination found annual net savings of $9.9 billion in direct costs and $43.3 billion in societal costs (Zhou et al. 2005). Several scholars have even suggested that some drugs be given away for free because they cut costs elsewhere (Choudhry et al. 2007).

Diagnostics, not just treatments, play a fundamental and growing role in reducing or avoiding health-care expenditures. In vitro diagnostics for HIV, hepatitis C, and other blood-borne infections have prevented the transmission of illness and bolstered the confidence with which physicians employ invasive procedures, such as blood transfusions. The ability to identify patients for whom drug therapies are likely to be effective reduces health-care costs and is sometimes the only way to make an innovative treatment feasible.

A case in point is Centoxin (nebacumab), which was developed to reduce mortality from sepsis, a dangerous and difficult-to-treat condition that has been estimated to kill more than 200,000 annually in the United States alone (Tanner 2004). Clinical trials revealed that Centoxin reduced mortality for patients under attack by Gram-negative bacteremia but increased mortality for other patients. Because no reliable diagnostic test for Gram-negative bacteremia was available, Centoxin was withdrawn in Europe. A very different situation surrounds Herceptin (trastuzumab), a monoclonal antibody used to treat advanced breast cancer. Herceptin only works for patients whose cancer expresses increased quantities of the HER-2 protein. Fortunately, Herceptin can be paired with a reliable test for the HER-2 protein—a drug-diagnostic combination that has become a fixture in breast cancer treatment (Danzon and Towse 2002).

Pharmaceuticals, including biotech drugs, have played a role in two spectacular trends of the past few decades. One is the striking decline in heart disease mortality by more than half since the 1960s. Looking beyond lifestyle changes such as reduced smoking, two recent studies ascribed most of the mortality decline to the development and use of new drugs (Ford et al. 2007; Weisfeldt and Zieman 2007). The second, equally remarkable, trend is a decline in disability rates among the elderly population, which have been dropping by roughly 1 percent per year (Cutler 2001; Freedman et al. 2002; Fries 2002). This has translated into a rapid decline in the nursing home population, despite the increasing number of elderly (Lakdawalla and Philipson 1999).

Using standard valuation methods derived from what people have been willing to pay to achieve marginal reductions in the risk of death, several economists recently estimated that recent advances in medical technology (of which pharmaceuticals are only part, although an extremely important part) are worth literally trillions of dollars (Murphy and Topel 1999).

Research Prospects

With biotechnology so new and its science and technology base expanding so rapidly, one can fairly say that in terms of health benefits, the field has barely emerged from its infancy.

An illustrative example is the sudden appearance of targeted cancer drugs, employing several very different mechanisms, after literally decades of basic and applied research that often seemed fruitless. This new generation of drugs includes Herceptin, Gleevec, Tarceva (erlotinim), Erbitux (cetuximab), and Avastin (bevacizumab). Avastin represents the first success in decades of research attempting to treat cancer through angiogenesis inhibition—that is, starving cancer cells by suppressing angiogenesis, the body's natural process of creating blood vessels to feed rapidly growing cancer cells. Despite its inherent promise, this research stream had met with decades of costly failure (American Cancer Society 2001; Barinaga 2000; *Wall Street Journal* 2004; and Marshall 2002). Now that trials have demonstrated the principle that angiogenesis inhibition can actually work, and Avastin itself has been approved to treat metastatic colorectal cancer, lung cancer, and breast cancer, vigorous new research is achieving encouraging results for other angiogenesis inhibitors. Writes Jean Marx (2005), "The strategy of denying growing tumors a blood supply continues to show clinical promise as new and improved drugs move through the pipeline." In the meantime, researchers are pursuing entirely new approaches to angiogenesis to improve both cancer and noncancer therapies (Jain 2005).

All of these drugs are being tested on stages or forms of cancer other than those to which they were originally targeted, generating a multitude of new treatments from the same drugs. Herceptin, for example, was originally approved for late-stage breast cancer, but it has proved extraordinarily effective as an adjuvant therapy after surgery at earlier stages. A recent *New England Journal of Medicine* editorial (Hortobagyi 2005) describing this line of research declared, "On the basis of these results, our care of patients with HER2-positive breast cancer must change today."[3] As discussed by Dooren (2005) and Calfee and DuPré (2006), the same thing has happened with three aromatase inhibitors: AstraZeneca's Arimidex (anastrozole), Pfizer's Aromasin (exemestane), and Novartis's Femara (letrozole). Gleevec, originally approved for one form of chronic myelogenous leukemia, has since been approved for a less advanced form, as well as for treatment of a rare intestinal cancer, while clinical trials on other indications continue (Calfee and DuPré 2006).

This new generation of cancer drugs has already provided remarkable benefits, as patients facing years of debilitating treatment have resumed relatively normal, even productive lives. A recent *Wall Street Journal* article described the new world of cancer survivors in unprecedented numbers (Marcus 2004). A *New England Journal of Medicine* editorial entitled, "Aromatase Inhibitors: A Triumph of Translational Oncology" (Swain 2005) captured the essence of the technology that makes this possible:

> Experts are now in the process of classifying breast cancer, which actually consists of a heterogeneous group of cancers, into multiple categories. It is essential to define each subgroup precisely and to delineate distinct characteristics and targets that will lead to tailored therapies that are better than the ones we have now. (2807)

The field of drug-diagnostic combinations is also making rapid progress. We have already mentioned Herceptin and the gene-based HER-2 protein diagnostic test. When Erbitux was approved to treat colorectal cancer, a gene-based diagnostic test was approved at the same time to identify patients whose cancer would respond to the drug (Mitchell 2004). Another *New England Journal of Medicine* article has reported on a close match between genetic test results and the ability of Iressa (gefitinib) to treat advanced lung cancer (Minna et al. 2004). There is every reason to expect dramatic benefits from pairing tightly targeted biotechnology cancer drugs with new diagnostics (Danzon and Towse 2002).

Many other examples of the early returns from biotechnology research and development could easily be cited. What is clear, however, is that the most important applications of biotechnology are still in their infancy. More than four hundred biotech drugs are in clinical trials involving more than two hundred diseases and conditions (Biotechnology Industry Organization 2007b). Biotechnology cancer drugs, as remarkable as they are, still deal with only a relatively small number of cancers, often restricted to certain genetic mutations; most remain to be developed. The pairing of genetic tests with biotechnology drugs has just begun, as most cancer drugs, for example, are still characterized by hit-or-miss successes with relatively small numbers of individual patients. Gene therapy, in which certain segments of patients' DNA are altered in order to prevent or cure disease, offers an immense potential for correcting genetic defects that is only beginning to be realized.[4]

A closely related area where research is also in its infancy is therapeutic vaccines—that is, agents that energize the immune system to attack pathogens or disorders that are normally ignored by the immune system (see Sela and Hilleman 2004 for a review). Late-stage clinical trials are underway for a range of illnesses including cancer (where examples include Dendreon's Provenge for prostate cancer and Cell Genesys's GVAX for prostate cancer) and Alzheimer's disease (including a stream of research at Wyeth and Elan that has generated tantalizing results in both mice and humans; see Schenk et al. 1999; Brendza et al. 2005; Nicoll et al. 2003; and Fox et al. 2005.[5]

Again, most of the fruits of biotechnology lie in the future, and we cannot know what they will be. As with all kinds of technological progress, advances in medical technology in general and in biotechnology in particular routinely defy prediction. Decades of frustration or failure are interrupted by sudden and striking success, often followed by new setbacks, with examples ranging from research on the role of thrombolytics (Brody 1995; Ross et al. 1999) and sepsis shock (Matthay 2001; Seiden 2001; *Datamonitor* 2002) to angiogenesis in cancer cells (American Cancer Society 2001; Barinaga 2000; *Wall Street Journal* 2004), the boosting of HDL ("good") cholesterol (Pearson 2006), and gene therapy.

But we can be certain of three things. First, the biotechnology applications already developed will be dwarfed in significance by the developments yet to come. Second, those developments will require the expenditure of many billions of dollars for basic and applied research. And, third, the size and effectiveness of those R&D investments, and the value of their results, will depend greatly on the institutional arrangements through which they take place, and, especially, the financial incentives to invest in the most productive of them.

2

The Sources of Biotechnology R&D

Biotechnology drug development is a striking example of the three-legged-stool model of technological development. Publicly funded basic research, much of it conducted in academia, triggers for-profit drug development, in a process that has become increasingly seamless and in turn often involves important feedback from industrial R&D to basic research.

Basic Research

The classic economic justifications for publicly supported research apply to basic research in biology and medicine. Such research can create results of great social value, yet the incentives for private parties to engage in it are weak. Most results of basic research can be easily replicated or exploited by public or private researchers. At the same time, many of the most important findings cannot be protected by patents or other forms of intellectual property (IP). This lack of protection drastically reduces the prospect of recouping research costs. Hence, the need for public subsidies.

Also a factor is the extraordinary unpredictability of basic research. Its serendipitous nature, with some of the most valuable results being totally unexpected, has been a prime justification for federal subsidies since publication of the landmark Vannevar Bush report in 1945:

> One of the peculiarities of basic science is the variety of paths which lead to productive advance. Many of the most important discoveries have come as a result of experiments undertaken with very different purposes in mind. Statistically it is certain

that important and highly useful discoveries will result from some fraction of the undertakings in basic science: but the results of any one particular investigation cannot be predicted with accuracy (Bush 1945).

This view remains valid.

The federal National Institutes of Health is by far the world's largest agency for conducting and funding basic medical and biological research. In real terms, the research budget of NIH burgeoned from $8.1 billion in 1990 to $26.4 billion in 2003, after which it increased, more modestly, to $28.6 billion in 2006 (American Association for the Advancement of Science 2005 and U.S. Department of Health and Human Services, National Institutes of Health 2007). Much of the research funded by this money, including the NIH's part in mapping the human genome (Venter et al. 2001), has proved extremely useful for biotechnology advances. A recent issue of *Science*, for example, highlighted three articles describing remarkable advances in delineating the genetic bases for age-related macular degeneration, an illness that often leads to blindness (Daiger 2005). Late in 2005, NIH embarked on construction of a "Cancer Genome Atlas" to investigate and catalogue genetic links with cancer (Pollack 2005).

Sometimes the distinctions between basic and applied biological research become indistinct. Advances in molecular biology have, at least moderately, reduced the role of serendipity in more basic research. As in the case of macular degeneration, applied genomics has demonstrated the ability to provide substantial leads toward finding useful inventions. At the same time, the 1980 Bayh-Dole Act has made it possible for researchers, including academic institutions, to obtain patents on NIH-funded research results.[1] This important development (discussed in more detail below) has partly overcome the lack of intellectual property protections and financial incentives necessary to exploit some of the practical results from basic research.

Notwithstanding these developments, however, the fundamental rationale for subsidizing basic research remains intact: The pursuit of intellectually challenging basic research sometimes yields social benefits or spillovers that greatly exceed potential profits from private investment.

Research in Academia

A large and diverse set of nongovernment institutions pursues the basic and applied research that forms the foundation for pharmaceutical innovation. Universities and stand-alone institutions conduct research in chemistry, biology, pharmacology, and allied fields ranging as far as particle physics (essential to imaging devices and their diagnostic capabilities) and computer software (which drives the apparatus of bioinformatics, genomics, and proteomics). Reports commissioned by the European Union have noted that the United States is unparalleled in the size, vigor, and independence of such research institutions, despite decades of efforts by EU nations to close the gap (Allansdottir et al. 2002; Gambardella, Orsenigo, and Pammolli 2000; Owen-Smith et al. 2002; Philipson 2005). Collaborations between publicly and privately funded R&D have accounted for some of the most valuable therapeutic breakthroughs in recent years. Reichert and Milne (2002), for example, examined the genesis of twenty-one exceptionally important drugs and described the intimate interplay of public and private research, along with similar details on the origins of other breakthrough pharmaceutical research (also see Stossel 2005).

Most biological and medical research in academia is supported by federal and, to a lesser extent, nonprofit sources (National Science Foundation 2005). Funding from pharmaceutical firms is also significant, however, and goes back at least as far as the years between World War I and World War II.[2] A survey by the National Research Council shows that in 1940, fifty U.S. companies were already supporting 270 biomedical research projects in seventy universities (Blumenthal 2003). Collaboration between universities and the pharmaceutical industry temporarily declined after World War II as federal support increased, only to surge again starting in the 1970s. By the late 1990s, over a quarter of the faculty members in life science departments at major U.S. research universities received research support from the industry (compared with 17 percent in 1983), while more than 50 percent consulted for it, and about 7 percent held equity in companies involved in work related to their own research. At the same time, over 90 percent of firms conducting research in life sciences were involved in research relationships with U.S. universities. In 1994, for instance, companies spent about $1.5 billion, spread over six thousand life science projects in U.S. universities, constituting about 14 percent of total funding for academic research in life sciences (Blumenthal 2003). Much of this activity involved patent licensing and associated royalties.

The United States is unique in the ease and richness with which scientific skill, research results, and practical applications flow from research institutions to pharmaceutical firms.[3] In 1980, only about 5 percent of federally owned patents were actually being used (Schacht 2000). Passage of the Bayh-Dole Act that year vastly expanded interactions among federal research agencies, academia, and industry. The number of patents issued to U.S. universities increased from 380 in 1980 to 3,151 in 1998 (Scherer 2002) and reached nearly 4,000 in 2003 and 2004 (Association of University Technology Managers 2005a; 2005b).

In biotechnology, some universities built patent portfolios rivaling those of major pharmaceutical firms. The University of California complex is the largest single holder of DNA patents, with the U.S. government in second place (Malakoff 2004). Universities maintained 69.6 percent equity in 450 start-up companies created in 2001, and in 2002 some 10,000 licenses yielded a gross income of $1.3 billion, more than $1 billion of that coming from royalties on product sales (Association of University Technology Managers 2002). In 2004, universities executed nearly 4,800 new licenses or options, most of them with small companies. More than 3,000 new products have been put on the market since 1998 (Association of University Technology Managers 2005b). All of this has been accompanied by increased industry funding for academic research, although as late as 1999, industry's share of total academic research funding had reached only 7.2 percent (from 4.1 percent in 1980; see Scherer 2002).

Private Sector Research

Since 1990, research expenditures by the pharmaceutical and biotechnology industries have increased in real terms from about $8 billion in 1990 to over $50 billion in 2005, substantially more than the entire NIH budget.[4] The NIH focuses on basic research, almost never moving on to the costly large-scale clinical trials required for Federal Drug Administration (FDA) approval of new drugs. Almost all new drugs are developed by private industry. An NIH study released in 2001 found that of forty-seven FDA-approved drugs with at least $500,000 in U.S. sales in 1999, only four involved direct or indirect federal patents (U.S. Department of Health and Human Services, National Institutes of Health 2001). In a review of comprehensive drug development databases, DiMasi, Hansen, and Grabowski (2003) found that of the 284 drugs approved in the United States during 1990–99, government sources accounted for 3.2 percent and academia 3.5 percent, with the other 93.3 percent coming from private industry. Reichert and Milne (2002) explored in more detail the relationships between public and private research, noting, among other things, the very limited extent to which the NIH engages in the kinds of clinical trials necessary to demonstrate the safety and efficacy of new drugs.

Some of the most fruitful industrial research has exhibited a combination of basic research and serendipity similar to that which often characterizes publicly supported research. One example, advancing through the laboratories of several independent firms, has been the long line of unexpected consequences from the discovery of estrogens in the 1930s. This research stream ran through the surprising discovery of the antiinflammatory properties of a few otherwise disappointing compounds to the development of the first Cox-2 inhibiting nonsteroidal antiinflammatory (NSAID) drug. Valuable new cancer and osteoporosis drugs were also generated as offshoots from this intricate and highly diversified collection of projects (Lednicer 2002). The recent announcement of promising results from a small, uncontrolled experiment in gene therapy for Alzheimer's disease illustrates, again, the intimate connection between practical cure-oriented research and basic proof-of-principle (Tuszynski et al. 2005).

Substantial basic research is located in the private sector. Data from the National Science Foundation (2007) indicate that about 14 percent of basic research is funded by private industry, compared to 8 percent funded by the federal government. Much of it is speculative and very much forward-looking in the sense of being conducted without a certain path to revenues. Indeed, although hundreds of biotech firms are collectively spending billions of dollars annually on research, the vast majority receive no drug revenues at all, and most of them never will. The extraordinary financial risks to this enterprise are discussed in more detail in the next chapter.

3

Essential Features of the Biotechnology Industry

Intellectual property regimes cannot work well unless they respect the essential characteristics of the industries in which intellectual property is important. The same is true of efforts to alter or reform the laws governing intellectual property; such changes may work poorly or even do harm if they ignore the nature of the markets in which they will take effect. This chapter describes the leading characteristics of the biotechnology industry.

Lengthy Product Development Times

Research and development in the biotechnology industry is notorious for its duration. This is understandable when one considers the origins of most biotechnological drugs. The research process often starts with the identification of a DNA sequence. Researchers then attempt to learn how the sequence functions, especially in the expression of proteins, and whether it shows signs of mutation with adverse health effects. Research may focus on apparent links among the proteins, gene mutations, and health disorders, raising the possibility of finding preventatives or cures by exploring relevant proteins, or perhaps even correcting mutations or flaws through gene therapy. Sometimes the process operates in reverse, beginning with the identification of a health problem or a suspect protein, and then working backward to identify the DNA sequence responsible for the disease (Scherer 2002). Work can then focus on exploiting this knowledge to isolate potentially useful proteins and related substances.

The second research stage translates what has been learned into a usable form, such as a protein-based drug, a diagnostic test, or a therapeutic modality. These products usually consist of one or more very large molecules, such as proteins, rather than the relatively simple chemical compounds or "small molecules" that long dominated the pharmaceutical industry. Then comes the arduous third stage, a long succession of preclinical and clinical tests. This work is similar to, but typically more complicated than, the clinical-trial process of traditional small-molecule pharmaceuticals.

Finally, during the fourth stage, the innovator seeks compliance with regulatory requirements before launching the final product onto the market. Because "biologics" (the FDA's classification for vaccines and most biotechnology drugs) are essentially grown rather than synthesized (as chemical-based drugs are), manufacturing them is far more complicated than manufacturing traditional drugs, and the regulatory apparatus is similarly complicated and time-consuming.

This description, while entirely typical, is misleading in its depiction of a one-way stream from basic research to the applied R&D that brings new drugs to market. Successful drugs often serve as essential tools for testing and revising basic hypotheses. The statin class of cholesterol-reducing drugs, for example, has forced a series of reassessments of scientific views not only of the role of LDL-cholesterol (so-called "bad" cholesterol), but also of the immediate cause of most heart attacks (Steinberg 2006; Topol 2004). Much the same process is now occurring in the wake of the successful deployment of angiogenesis-inhibitors (Ferrara 2002) and TNF-inhibitors (such as Rituxan, approved for treating both rheumatoid arthritis and cancer).

The relationship between the biotechnology and traditional pharmaceutical industries is complex. While both transform basic research findings into medical treatments, biotechnology products and the means for producing them are typically, as noted, vastly more complicated. In addition, the tools of biotechnology, ranging from genomic sequencing to such complex hardware-software combinations as gene arrays, have been transforming traditional small-molecule R&D. The later stages of product development in the pharmaceutical and the biotechnology industries are now quite similar and often intermingle. In both, preclinical and clinical trials are almost always long and expensive, and the complex formalities of regulatory approval always precede the launch of the final product on the market.

But, again, biotechnology manufacturing poses unique challenges and expenses, not least of which are the FDA's stringent requirements for facilities. Designing and building factories for biologics is expensive to begin with, mainly because, as already mentioned, these drugs essentially must be grown rather than synthesized in a purified form, as are traditional, "small-molecule" drugs. In addition, the FDA typically requires manufacturers to construct full-scale factories to produce the drugs used in the clinical trials that will determine whether the drugs will even gain marketing approval. This raises costs beyond those in traditional pharmaceutical R&D, where the FDA is satisfied when large-scale manufacturing facilities can be demonstrated to produce "bioequivalent" products (something that is usually almost impossible for biologics).

Other practical differences between the two industries are also fundamental. Most biotechnology firms are small in comparison to their pharmaceutical counterparts, and they lack the internal financing resources necessary to undertake drug development. Many carry out what amounts to basic research to identify potential new products, and then enter into partnerships with larger pharmaceutical companies for testing and development. Patents facilitate this process, leading to extensive licensing between the two industries (U.S. Federal Trade Commission 2003). With capital markets in turmoil in recent years, some biotechnology firms aspire to become part of fully integrated pharmaceutical companies, while others subscribe to the growing trend of partnerships and mergers among participants in the biotechnology industry (Ernst and Young 2003).

Research on drug development in the traditional pharmaceutical industry indicates that it takes an average of nearly eight years to move from the start of clinical trials through FDA approval. (It took nearly a year longer until the Prescription Drug User Fee Act of 1992[1] helped reduce new drug review times at the FDA; see DiMasi, Hansen, and Grabowski 2003.) Preclinical research can add years to this timeline. Biotechnology drugs generally require even more time than small-molecule drugs. For instance, the Immunex Corporation—a U.S. biotechnology company founded in 1981—needed ten years to bring its first product to the market, and another six to bring out its second (U.S. Federal Trade Commission 2002a).

Obtaining final marketing approval from regulators can itself require a long time. Since 2000, approval times for U.S. Food and Drug Administration reviews of new biotech drugs have averaged between 12 and 19 months (Grabowski 2007, 33). During 2002, the FDA approved twenty-six new drugs in an average time of 17.8 months (Pharmaceutical Research and Manufacturers of America 2003). As described below, one result of this lengthy process is to reduce the effective life of biotechnology drug patents to substantially less than the theoretical maximum of twenty years from the date when a patent is filed with the U.S. Patent and Trademark Office (USPTO).

The Costs and Uncertainty of Bringing New Drugs to Market

Quite aside from the time it requires, biotechnology drug development is expensive and fraught with uncertainty. The most comprehensive data pertain to traditional drug development. R&D costs have escalated in the past two decades, as mandated clinical trials have become larger and more complex and safety margins have increased. An analysis of the R&D costs for sixty-eight randomly selected drugs (including six biotechnology drugs) developed in-house by twelve pharmaceutical firms during the 1990s found that the average cost of bringing a new drug to market was $802 million in 2000 dollars (DiMasi, Hansen, and Grabowski 2003; also see Dickson and Gagnon 2004). This sum takes into account the time value of invested funds and the amount spent on products that never made it to market. Looking back on several decades of research on drug development costs, the authors concluded that they are rising at a compound rate of about 7 percent above inflation. Economists at the Federal Trade Commission (FTC), using very different methods, recently estimated slightly higher costs (Adams and Brantner 2004). Another recent report updated the DiMasi, Hansen, and Grabowski study to use 2004 dollars, while adjusting for higher launch costs and the apparent slowdown in approvals of new drugs. It estimated average costs at about $1.7 billion per new drug (Gilbert, Henske, and Singh 2003).

This rapid escalation arises primarily from the costs of clinical trials. Since the first careful study of development costs was published in 1979, clinical costs have grown roughly five times faster than preclinical, calculated on an inflation-adjusted basis. The growing share of clinical costs in R&D expenditures reflects the difficulty of recruiting patients into clinical trials (in an era of expanding scale in drug development), a shift toward drugs to be taken by large numbers of patients with chronic and degenerative diseases, and a growing emphasis on safety, which have combined to cause a sharp increase in the number of procedures per patient in clinical trials (DiMasi et al. 2003, 162, 167; on procedures per patient, see DiMasi et al. 2003, 162n21).

This research also documents the extraordinary uncertainty involved in testing a promising molecule or biological for marketing approval. Only about a third of approved drugs ever produce revenues sufficient to cover full development costs. The rest either fare disappointingly in the marketplace, or they do no more than cover the costs of the final series of clinical trials that were launched after earlier costs had already been absorbed by the firm (Grabowski and Vernon 2000). Some of the most promising drugs—such as torceptrapib, Pfizer's HDL-cholesterol booster—have failed after investments that approached one billion dollars (Pearson 2006).

Even for approved drugs, revenue predictions often prove unreliable when market realities intrude (Spilker 1994). Truly striking innovations to treat fatal conditions that had defied treatment despite decades of research may fail to yield anything like the profits initially expected by market observers. A relevant example is Xigris, Eli Lilly's remarkable recombinant human-activated protein C for the treatment of severe sepsis. A market research report published about the time Xigris was approved (and before the drug was renamed) carried the title, "Zovant (for Sepsis)—A Potential Blockbuster" (Seiden 2001). A year or so later, another market research report was published: "Xigris: Lilly's Sepsis Flop Misses Out" (*Datamonitor* 2002). Sales had rapidly leveled off (Regalado 2003), and Lilly inaugurated costly new trials to provide more thorough documentation of the benefits of its remarkable drug.

There are several reasons to think that, on the whole, the uncertainty and costs of the biotechnology R&D process are at least comparable to those for pharmaceutical R&D generally, and may be even greater. Grabowski (2002, 92–95) described the relatively rapid development of some early biotech drugs, where the goal was to create recombinant versions of naturally occurring proteins whose functions were well understood, in contrast to the steady increase in development times as the industry tackled more difficult problems. Always, the construction of manufacturing facilities can involve substantial delay. Biologics manufacturing can itself require substantial research and development, and most of these problems must be solved to produce the quantities needed for clinical trials, while satisfying the FDA about the integrity of what is often an organic combination of a biological product and a manufacturing facility. In the rare situations where a product is approved before manufacturing facilities are constructed—as for some vaccines—the task of building and validating manufacturing facilities can take as long as five years (Wess 2005).

Some of the most promising research lines fail many times before the first success is achieved—if, indeed, any victories at all are forthcoming. We mentioned the decades-long search for a way to harness protein C to treat severe sepsis (which continues because Xigris, the first fruit of that search, often fails). A potentially more momentous story is research on gene therapy, in which certain segments of patients' DNA are altered to prevent or cure disease. Gene therapy offers immense potential for correcting genetic defects. Tantalizing animal research has revealed a future fraught with scientific challenges, as well as extraordinary promise (High 2005; also see Ball and Anderson 2000 for prospects a half-decade earlier). Recent work on zinc finger nucleases (a class of proteins) offers hope on the most vexing problem, the design of vectors to permit gene-correction mechanisms to reach each of the billions of cells to be fixed (Kaiser 2005).

But after the investment of billions of dollars in some two decades of research, the FDA has yet to approve a single gene therapy for human use. The CEO of a biotechnology firm was quoted in the February 18, 2005, *Wall Street Journal* as saying, "Gene therapy has been 'five years away' for 20 years" (Begley 2005). Indeed, an August 29, 1999, *New York Times* article noted that an apparently promising gene therapy was "at least three to five years away from the market" (Fisher 1999).

Nonetheless, work proceeds. An experimental but dangerous treatment has been developed for an extremely rare immunodeficiency condition (Rosen 2002). A small trial in delaying the progression of Alzheimer's disease has shown great potential (Tuszynski et al. 2005). Also promising is research on treating Parkinson's disease through gene therapy after traditional drug administration has failed, and on treating sickle cell anemia.[2] But, again, the prospect for substantial revenues remains hazy at best.

Another example of the uncertainties of biotech drug development is the already mentioned attempt to control or cure cancer through angiogenesis inhibition—that is, by cutting off the blood supply to rapidly growing cancer cells. Again, a long and costly series of failures preceded the first success. Not until 2004, after three decades of research, was the first drug employing angiogenesis inhibition approved by the FDA.

We have noted the extraordinary research underway on therapeutic vaccines and gene therapy. Promising results have been reported for cancer, Alzheimer's disease, Parkinson's, and other conditions. So far, however, we have seen mainly dashed hopes rather than usable products. Such disappointments are to be expected when trying to bring to market products that rely on mechanisms which, for the most part, have never been used in trials with human subjects. Nonetheless, work continues on many fronts.[3]

Competition among Biotechnology Drugs

For the biotechnology industry, competition occurs in two ways. One is between drug therapy and alternative treatments. When new cancer drugs are approved, the medical community may see them as competing with existing chemotherapy or even surgery. This tends to influence pricing, perhaps acting as an informal ceiling, even though drugs created through biotechnology methods (such as Gleevec or Herceptin) often save health-care and workplace costs because use of them is much less burdensome on both patients and health-care providers than use of other treatments (Vrazo 2005).

Nonetheless, the medical community is very aware of the financial trade-offs between the costs of biotechnology treatments and older approaches (as illustrated by Neyt, Albrecht, and Cocquyt's 2006 study on the cost-effectiveness of Herceptin). Some new drug prices are explicitly based on the costs they permit health-care facilities to avoid, as in the case of Revlimid (lenalidomide; Celgene), which is used to treat a rare blood disease that normally requires numerous transfusions (Anand 2005).

Most competition, however, involves other new drugs. Most people assume that real competition comes only with the arrival of generic drugs after patents expire. Economic research has found, however, that an equally important competitive force comes into play long before this: competition among follow-on drugs—that is, drugs that employ mechanisms similar to that used in a pioneering drug but do not involve the exact same molecule. When a firm invents a pioneering drug that creates a new therapeutic category, and that drug achieves success in the marketplace (which is by no means inevitable), one effect is the incentive for competing firms to enter the new market by developing follow-on drugs. This kind of dynamic has assumed great importance in high-technology markets generally. A study of commercial innovation in the United States found that the average time for entry by competitors to a pioneering innovation declined from 32.75 years in the period 1887–1906 to 3.40 years in 1967–86 (DiMasi and Paquette 2004, fig. 1, using data from Agarwal and Gort 2001).

The modern era of pharmaceutical development reveals a similar trend. In an exhaustive survey of new drug approvals over several decades, DiMasi and Paquette (2004) found that under the impact of competition and new technology, the period of exclusivity for pioneering drugs decreased from 10 years in the 1970s to 1.2 years in the late 1990s, with about one-third of follow-on drugs receiving priority review at the FDA, and that competitive entry has been accelerating at the rate of two to four years per decade since the 1960s. They also found that approximately one-third of follow-on drugs received a "priority" rating from the FDA when they were submitted for new drug approval during the years they studied, and that 57 percent of all therapeutic classes saw at least one follow-on drug with a priority rating. Lichtenberg and Philipson (2002), examining the market impact of follow-on drugs, estimated that competition from them actually reduced the total market value of a pioneering brand at least as much as the advent of generic competition after patent expiration.

Although these results pertain mainly to traditional pharmaceuticals, it is clear that the biotechnology industry has entered an era of vigorous follow-on competition as well. The FDA, for example, recently approved Bristol-Myers Squibb's Orencia, a TNF-inhibitor that is effective against rheumatoid arthritis, a disabling condition that largely resisted treatment until the advent of biotechnology. Orencia is the fourth TNF-inhibitor to reach market. It will compete with earlier entrants while also offering treatment to the estimated 15–25 percent of rheumatoid arthritis patients who do not respond to existing drugs (Saul 2005). Competition is also strong among drugs to treat schizophrenia (Abboud 2004), colorectal cancer (Greil 2005), multiple sclerosis (where, for example, Biogen-Idec and Serano have used different growth media to produce competing biological drugs; Krasner 2004), and breast cancer (for which three members of the new class of aromatase inhibitors are providing strong competition to older, more toxic chemotherapy; Zimmerman and Hensley 2004 and Dooren 2005). The development of DNA-based diagnostics and treatments is an extraordinary new twist to this long-standing form of competition.

Thus, we see the essential elements of all pharmaceutical competition: new products with superior side-effects profiles, easier administration, superior efficacy (often for specific subpopulations), and, often, vigorous price competition (DiMasi and Paquette 2004). We also observe that this entire process is facilitated by a central feature of the patent system: the requirement that inventors publish their work, thus easing the task of competitive drug development.

4

The Role of Intellectual Property Rights

Intellectual property protections, mainly patents, undergird biotechnology drug development. By providing the prospect of exclusive marketing of new diagnostics, tools, and therapies, they motivate the investment necessary to fund what is usually a long and expensive development process. But patent law itself is a complex, controversial, and rapidly evolving mechanism.

The Foundations of Intellectual Property

Article 8 of the U.S. Constitution provides that Congress shall have the power "to promote the Progress of Science and useful Arts, by securing for limited Times to Authors and Inventors the exclusive Right to their respective Writings and Discoveries." This incorporates what is sometimes called the utilitarian approach to intellectual property rights, in which incentives to create new inventions are balanced against the benefits of relatively unrestricted public access to and use of inventions after a reasonable period of time has passed. The American approach was essentially unique when the Constitution was written, but all advanced nations have since adopted similar provisions to protect inventors from uncompensated use of their inventions.

From an economic standpoint, intellectual property protections, including patents, embody a balance between two extremes, each of which would hobble technological advance.[1] One would be to provide no property protections at all to inventors. Sometimes, secrecy can provide reasonable protection against copying of inventions by competitors, but this is often impossible in industries where regulators and users of new technology require detailed information about the products they endorse or use. Because nonsecret inventions typically can be copied at low cost (at least

24

relative to the costs of discovery and development), competitors would quickly force market prices down toward the marginal costs of manufacturing. This would eliminate most of the expected profits that could compensate inventors for the costs and financial risks inherent to the inventive process, thus removing the most important incentive to undertake costly R&D. This "dynamic inefficiency" (so-called because it plays out over time) would inhibit innovation, so that many of the most valuable advances would be greatly delayed or perhaps not made at all.

The other extreme would be to provide inventors with permanent protection against appropriation of their inventions by competitors. Patents and other intellectual property protections generate prices well above marginal costs, however. This "static inefficiency" would impede usage by buyers for whom the product is worth more than manufacturing and distribution costs but less than market prices. Perpetual patents would usually keep prices well above marginal costs until close substitutes could be brought to market, a process that might take many years.

The universally accepted compromise avoids the two extremes through two tools. First, inventors are granted patent rights for a limited period of time, recently set in the United States and other advanced nations at twenty years from the time a patent application is filed. Second, inventors are required to disclose publicly the essentials of their inventions. This greatly facilitates the development of competing products while contributing to advances in basic and applied science. To make this system work, patent-granting authorities must make reasonable decisions when they determine whether applications meet the statutory requirements for novelty, nonobviousness, and practical utility.[2]

Why Intellectual Property Protection Is Necessary in Biotechnology

While the basic economic principles and legal rules just outlined apply to all industries, their impact varies greatly from one to another, reflecting differences in science, technology, and market conditions. Even within a single industry, the functions of patents and other intellectual property can change greatly over time and according to specific circumstances. For example, Mazzoleni and Nelson (1998) and others have pointed out that patents can motivate new inventions, and/or the commercialization of inventions that might have been created (but not commercialized) without patent protection, and/or motivate the efficient coordination of follow-up research.

We will address recent developments in the economics of patents in chapter 5, but a few points are essential here. As the central features of biotechnology described in chapter 3 clearly imply, there are few, if any, industries to which intellectual property protection is as important as it is to pharmaceuticals generally and to biotechnology in particular. In striking contrast to almost all other large industries, upfront sunk costs comprise some 70 percent of drug costs, with manufacturing and other short-run costs accounting for only about 30 percent (although manufacturing costs for some biotech drugs, such as monoclonal antibodies, can be much higher than for traditional drugs; Danzon 1998, 295–97). The industry will not remain viable unless revenues greatly exceed the costs of drugs actually brought to market and compensate for financial risks associated with the numerous research failures that yield no marketable drugs at all.

These conditions ensure that without intellectual property protections, imitators can easily undercut the prices of innovative new drugs. Secrecy is, for the most part, not a viable alternative to intellectual property. The FDA and international regulatory authorities, such as the European Union's European Agency for the Evaluation of Medicinal Products (EMEA), require detailed disclosure of almost all relevant intellectual property. Much disclosure is also necessary to attract and maintain the confidence of investors and, ultimately, the enduring confidence of physicians, patients, and health-care providers, without which success in the marketplace is impossible.

The Value of Patents to Innovation

A small but important literature has explored the impact of patents on innovation.[3] With few exceptions, demonstrating a positive relationship between the two has proved difficult. The leading exception, however, is pharmaceuticals,[4] as two decades of economic research have found. A widely cited study by Mansfield (1986) found that chemical and pharmaceutical firms regarded patents as important competitive tools and claimed that 65 percent of innovations in the pharmaceutical industry and 35 percent in the chemical industry would not have been brought to the market without them. No other industry showed more than 18 percent of newly launched products dependent on patent protection. While a 1989 review of the economic literature on innovation and market structure by Cohen and Levin concluded that firms in more than a hundred industries considered patents less important than trade secrecy, early entry, and customer service for competing in new product markets, knowledge-based industries—including chemicals and pharmaceuticals—again identified patents as crucial to their competitive positions. Two more recent surveys suggested the continuing importance of patents in the biotechnology, pharmaceutical, and chemical industries, despite an increasing emphasis on secrecy and nondisclosure in others (Johnson, Cohen, and Junker 1999; and Cohen, Nelson, and Walsh 2000).

Patents are especially crucial to the innovative R&D mounted by small startup biotechnology firms. Indeed, they are typically the only assets those firms possess that are sufficiently stable and valuable to attract the large amounts of capital they need to exploit promising research toward new drugs and diagnostics. Hence, market valuation of startup biotechnology firms tends to reflect the scope and breadth of their patents. One study, for example, examined a sample of 535 financing rounds for 173 private biotechnology startups in 1978 and 1992. It found that "firm value rises with the number of patents and the breadth of intellectual property protection, . . . [so that] a one standard deviation increase in average patent scope leads to a 21 percent increase in the firm's valuation" (Lerner 1994; the patent scope variable was defined as a function of the number of patent office technological classifications under which the patent was issued). Another investigation, based on a sample of more than six hundred original patents awarded to twenty leading biotechnology firms, estimated the market value of individual biotechnology patents as a function of patent content (Austin 2000). The study revealed market valuations of between $9 million and $14 million each for "protein" patents in leading research areas such as erythropoietin, colony-stimulating factors, human growth hormone, and hepatitis-B vaccine. The study also found that patents for innovations in recombinant DNA or genetic engineering were valued at between $13 million and $21 million. A more recent report, which focused on technology exchange among firms, found that stronger patents generated higher market value for firms (Arora, Fosfuri, and Gambardella 2001).[5]

Some of this research illuminates another important phenomenon: the tendency for biotechnology patents to yield positive spillover effects for the market at large. Approximating the extent to which one firm's intellectual property benefits competitors is quite difficult. Austin (2000) attempted to do so, however. He estimated that patent disclosure generated net public-knowledge spillovers to the industry with values of between $2 million and $4 million per firm per rival company, and that patents for innovations in recombinant DNA or genetic engineering yielded $3 million to $6 million per rival company. This study revealed that patents are not simply winner-take-all prizes. On the contrary, individual patents can provide a direct route to knowledge spillovers, providing economic value for both the patentees and their rivals. The work by Arora et al. found that stronger patents made for more efficient technology licenses, even for unpatented technology.[6]

Also of great importance, but apparently unmeasured, is the tendency for successful drugs to serve as research tools for testing, refining, and suggesting hypotheses in basic science. This has happened, for example, in cardiology, where large-scale clinical trials of cardiac stents and cholesterol-reducing drugs have forced the reexamination of basic hypotheses (Steinberg 2006; Calfee 2007b). Patent protection is necessary to motivate most of these trials.

Unfortunately, research has also documented the tendency for the U.S. regulatory system to limit biotechnology's contribution to the innovation process through the steady erosion of patent life by testing and other requirements. The nature of biotechnology drug development greatly reduces effective patent length from its maximum twenty years from time of filing. A decade or more can be occupied in testing, in constructing and validating manufacturing facilities, and in otherwise dealing with FDA and other regulatory oversight and review. The reduction in time experienced in traditional small-molecule drug development shows the problem clearly. Traditional pharmaceuticals launched in the mid-1990s were, on average, protected by less than twelve years of effective patent life, with a maximum period of fourteen years after FDA approval, despite occasional extensions (Grabowski 2002, 100; also see Desrosiers 1989). In contrast, nonpharmaceutical, knowledge-based products that do not require regulatory approval typically enjoy more than eighteen and a half years of effective patent life (Pharmaceutical Research and Manufacturers of America 2002).

The Evolving Economic Rationale for Biotechnology Patents

Over the past decade, a flood of new research has advanced—and sometimes complicated—our views of the benefits and costs of the patent system. We have noted the general utilitarian theory of patents, which posits the necessity of intellectual property protection to overcome market failure in the form of so-called "free riding" by competitors making unauthorized use of the patentee's innovation. Under this "canonical" version of patent theory, the patent system is deemed indispensable in motivating useful invention, which would occur far less frequently without it (Mazzoleni and Nelson 1998). Branching out from this initial explanation of the patent system, more recent work has stressed complementary theories about its principal purposes and effects.

In general, while government officials, the business community, and academic researchers have become increasingly aware of these complexities, opposition to moving from a unitary system with a single set of laws and regulations for all industries toward a multiple system tailored to the characteristics and imperatives of individual sectors remains strong.

Where, then, does biotechnology fit in this more nuanced theoretical framework? To answer this question, one needs to explore several recent explanations of the working of the patent system. For instance, it has been pointed out that in addition to motivating innovation, patents also serve a societal purpose of encouraging disclosure and thus providing a means for the wide diffusion of technological information. Accordingly, the inventor gains additional returns through the use of the product or process by others. This insight assumes that the inventor alone cannot exploit all of the uses of the invention, thus motivating extensive cross-licensing with other inventors or companies. The activity of patenting around successful inventions is often seen as another benefit from the diffusion of information.[7]

A second theory posits that a key function of patents is to enable inventors to approach capital markets to compete for development financing, or to combine with larger entities to attract the investments necessary to commercialize inventions (Eisenberg 1997). This insight has direct relevance to biotechnology for several reasons. First, many biotechnology firms are, in fact, quite small, lacking the resources to commercialize the process or product patent they obtain. A common occurrence is for the original company to link up with a larger company, usually in the pharmaceutical industry. In effect, these small firms' capital consists almost entirely of their intellectual property. Second, while not explicitly set forth at the time, the capital markets theory was central to the passage of the Bayh-Dole Act in 1980, which, among other things, gave universities patent rights on products and processes that evolved out of government-funded projects. The argument underpinning Bayh-Dole was that patents were needed to induce commercialization of the results of public research by the private sector, which it would not otherwise finance without the limited monopoly derived from them. In the two decades since the passage of Bayh-Dole, universities have become big players in the patenting game, with some of the most lucrative patent alliances with industry coming in biotechnology (U.S. Federal Trade Commission 2003).

Looking ahead to later sections of this study, it should be noted that in their testimony regarding proposed reforms of the U.S. patent system, biotechnology firms have stressed investor confidence and support as the central factor behind their defense of strong patents. One knowledgeable commentator described the biotechnology industry's argument as follows:

> Investors believe that in order for the biotechnology sector to succeed, it is critical that biotechnology firms be able to obtain and enforce strong patents. Biotechnology companies, particularly those that have yet to put a product on the market, must rely on substantial investment funding in order to survive. If there is any perception that patent reform will weaken patent protection for biotechnology inventions, investors will not be as willing to fund biotechnology and this reluctance will adversely impact biotechnology. (Holman 2006, 327–28; see also testimony of BIO on 2007 patent reform legislation in U.S. Senate 2007a; U.S House of Representatives 2007a)

Finally, an insight that was first articulated in the 1970s and achieved new prominence in the 1990s has prompted wide discussion and debate in recent years. It is labeled the "prospect" theory of patents and was advanced originally by law professor Edmund Kitch in a seminal 1977 article (Kitch 1977; Burk and Lemley 2003). Aiming to integrate intellectual property rights with more general property rights theory, Kitch argued that the IP system is best viewed not so much as an incentive-reward system, but rather as a prospect system, analogous to a mineral claims system. In this view, granting a broad patent allows for the development of a full range of economic possibilities in a reasonably efficient fashion, avoiding inefficient duplication. The single patentee is best able to coordinate the development and/or improvement of an invention. The prospect theory also aims to overcome the so-called "tragedy of the commons," which describes situations where the social incentive to "invest" in the long-term value of a lake or a field—for example, by letting it lie fallow, or limiting the grazing or fishing to permit naturally occurring forces of renewal to maintain the property's productivity—is less than the private value of individual exploitation (Hardin 1968). Without some kind of oversight, the field or lake will be overfished or overplanted, thus reducing both public and private welfare. In Kitch's terms, a broad patent is necessary to avoid "wasteful mining of the prospect."

In addition to these new theories regarding the role of patents, certain conditions of competition and market structure need to be discussed in evaluating potential challenges to innovation in biotechnology. The first set of issues revolves around the differentiated nature of innovation in individual sectors (Burk and Lemley 2003). The original incentive-rewards theory of patents assumes "stand-alone" innovation, or a stylized model involving a single invention. The traditional pharmaceutical industry has been identified as a quintessential model for this type of innovation—in part because some of its most significant innovation is discrete, although it can also provide an essential foundation for subsequent discoveries of additional molecules. The industry is also characterized by a lengthy and expensive development process (with its clinical trials and regulatory requirements), a high likelihood of failure, and the relative ease of imitating patented processes or products and free riding on their coattails.

In recent years, however, researchers have become interested in models of innovation that do not fit the stand-alone pattern evinced in some sectors. They have pointed to the importance of cumulative innovation, with an attendant (though separate) potential problem of an "anticommons" (Scotchmer 1991; Heller and Eisenberg 1998; Burk and Lemley 2003).

Theories describing cumulative innovation see innovation as an ongoing, iterative process in which many contributors build upon the work (the "shoulders") of each other. In sectors where innovation is cumulative, one inventor alone cannot reap the crucial gains and must depend on others to advance and optimize a product line over time. Inevitably, this raises issues of dividing property entitlements to maximize incentives for cumulative improvements. The software industry is often cited as a model for cumulative innovation, but innovation in the biotechnology sector is also increasingly characterized by cumulative research extending over a considerable number of years (see below; also Calfee and DuPré 2006).

Linked to the model of cumulative innovation are alleged problems related to the "anticommons" (Heller 1998; Heller and Eisenberg 1998). A stream of literature has pointed out the possibility that granting too many patents will fragment property rights and inhibit innovation. The argument is that too many companies or patent-holders may be granted too many patents on inputs or components of a final product, a situation that can lead to high transaction costs (in negotiating licenses) and strategic behavior (specifically, by holdouts who will charge exorbitant fees for the inputs or components).

In assessing where these theories and conditions apply to biotechnology, it is important to distinguish it from the closely related pharmaceutical industry. As noted above, biotechnology focuses on cells and large biological molecules (DNA and proteins) rather than the chemical compounds from which the pharmaceutical industry constructs small-molecule drugs. Although the biotechnology industry has many facets, generally its products can be classified into two types of inventions: newly discovered and isolated genes or proteins, along with pharmaceutical drugs based on them; and inventions that produce new methods of diagnosing and treating patients with particular diseases through the use of these genes and proteins (Biotechnology Industry Organization 2005; U.S. Federal Trade Commission 2003).

Biotechnology and pharmaceuticals share a number of characteristics as well. Both must endure long and expensive development and testing times and jump high regulatory hurdles, and then face situations where it is relatively easy for imitators to reproduce their drugs at substantially or even radically lower costs, and with much less uncertainty. Strong synergies exist between the two industries in attacking human diseases. Both are trying to produce end-use products, but the discovery of traditional pharmaceutical (small-molecule) drugs has been accelerated by the use of biotechnology tools such as proteins and genomic sequences. This has led to close structural relationships in many cases, with biotech companies often forging ahead with the research and then partnering with larger pharmaceutical firms to commercialize products. Cross-licensing and mergers are rampant between the two (U.S. Federal Trade Commission 2003; Ernst and Young 2007).[8]

But the biotech and pharma industries differ in important ways. As noted, many biotech firms are quite small and often lack the financial resources to tackle commercialization—in many cases, they are essentially collections of intellectual property rights waiting to be exploited. In addition, much biotech research is basic research, some steps removed from the applied research than can readily be translated into end-use products. An example is the discovery, isolation, and exploitation of the vascular epidermal growth factor by Genentech scientists while exploring the phenomenon of angiogenesis, the process by which cancer cells (among others) generate new blood vessels in order to grow. This supported a burst of basic research while also leading to the development of successful new treatments, including Avastin for cancer and Lucentis for blindness caused by age-related macular degeneration (Ferrara 2002).

Finally, many patents in biotechnology are for research tools and diagnostics, products that pose quite different challenges to patent theories (Landes and Posner 2003). Research tools and diagnostics provide the underpinning for discoveries from follow-on R&D, raising the danger that the creation of an anticommons could impede advances from basic research to marketable products. Furthermore, the cumulative nature of some advances in biotechnology poses difficult questions with regard to balancing rewards and incentives among those who conduct the initial research and those who engage in follow-on research. As Suzanne Scotchmer, a leading authority on cumulative innovation, has noted,

> The problem arises [from the fact that] the earlier innovators are laying a foundation for later innovators. And . . . in a sense they're creating an option on later innovations. That option has value. How do you reward the earlier innovators for the option they create for later innovations? (U.S. Federal Trade Commission 2002b, 135)

On the other hand, she adds, there is also a danger from overprotecting the original inventor by granting him a broad patent "that can stifle follow-on discovery . . . and if you stifle the follow-on, you also stifle the prior innovation and the whole research line dies" (135).

Clearly, issues remain unresolved about how to induce optimum efficiencies in the innovation process, given the highly idiosyncratic nature of individual technology sectors. Proponents of the prospect theory argue for broad early patents that will allow an orderly, efficient development. In their view, multiple, narrow patents will result in wasteful patent races that will sap the incentives for individual inventors to carry to conclusion the complex, expensive, and hazardous process of drug development. They find some support in at least some writings of those who espouse the cumulation theory with reference to innovation in biotechnology (Scotchmer 1991; Green and Scotchmer 1995). These researchers worry that the original inventor will not be able to share the returns from follow-on inventing, weakening the incentives to make the initial investment in pioneering inventions.

Arrayed against these views are those of a number of experts from both academia and the business sector who argue that patent races, on balance, will have positive results because different inventors will contribute from diverse perspectives, and ultimately this competition of ideas will advance innovation more rapidly and broadly (Mazzoleni and Nelson 1998; Merges and Nelson 1990). These skeptics also express great doubts about the ease with which licenses will be granted among potential competitors (given the inscrutable nature of the universe of possible follow-on inventions), a key factor in the benign paradigm posited by prospect and cumulation advocates.[9]

In the meantime, biotechnology drug development has proceeded in a diversity of ways compatible with competing theories of patents and R&D. Consistent with the prospect theory of patents are the length and richness of R&D agendas arising from narrow sources. An example is Genentech's Avastin. It was created to target a protein discovered in 1993, began clinical trials in 1997, was approved in 2004 for colorectal cancer, has since been approved for lung cancer, and is in testing for a score or so more cancers, even as the closely related drug Lucentis was developed in parallel to become approved as the first effective treatment for age-related macular degeneration. Comparably lengthy research agendas have brought a stream of new uses for such other drugs as Herceptin, first approved for late-stage breast cancer but demonstrated years later to be more effective for early-stage breast cancer. At the same time, however, it has become increasingly apparent that the tools of biotechnology open the door to rapid "inventing around" of many of the most innovative drugs (that is, firms can invent another drug that attacks the same biological mechanism but does not infringe upon the patents of the pioneer drug). Follow-on drugs that have been approved or are in late-stage testing exploit essentially the same biological mechanisms as those targeted by Avastin, Herceptin, and other notable pioneer drugs, such as Gleevec (Calfee and DuPré 2006; Calfee 2007a).

This chapter has attempted to lay out the most important theories, both complementary and conflicting, that are exerting powerful crosscurrents on the achievement of optimum intellectual property and competition public policies for the biotechnology industry. There are no easy answers to the challenges posed here; indeed, the prime conclusion is that at present there is not enough empirical evidence or historical experience to determine conclusively which of the contending theories and assumptions is more nearly correct. Two things are clear, though: First, the biotechnological research enterprise itself is remarkably robust in the face of various potential patent-based barriers to innovation; and, second, given the indeterminacy surrounding these issues and their practical consequences, caution should be the watchword for public policy in the face of proposals to make drastic changes to the current system.

5

Challenges to the Biotechnology Property Rights System

The following public policy analysis and recommendations will deal with two broad (and to some degree overlapping) clusters of issues. First, a number of problems have emerged recently regarding biotechnology patents and the innovation process, and the roles of the universities and public agencies that support the U.S. research and innovation enterprise. Of great importance here are questions concerning the consequences (intended and unintended) of the 1980 Bayh-Dole Act; the so-called upstream patents in biotechnology that potentially affect both subsequent research activities and commercial exploitation; and the wider claims of the deleterious impact of a growing "anticommons" phenomenon.[1]

Second, over the past two or three years, momentum has gathered for comprehensive and sweeping reform of the patent system, and a number of bills have been introduced that have major implications for biotechnology innovation and commercialization. Much of the impetus for change stems from flaws perceived in the institutions (the patent office and the courts) that administer the U.S. patent system.

In this chapter, we will describe and evaluate these issues and claims and advance a set of recommendations for what should be included—and not included—in future legislation. The major theme will be, "First, do no harm."

Property Rights and the U.S. Research Enterprise

Notwithstanding the obvious role of patents in motivating investment in R&D, many intellectual property experts emphasize the importance of preventing them from interfering with scientific research (Nelson 2003). The industry features described above define biotechnology as a sector with an extended research process that incorporates complex sequences of multiple discoveries often performed by separate, distinct entities. Therefore, patents have the potential to inhibit follow-on innovation, either when multiple property rights' owners claim rights to various inputs of research, or when patent owners simply withhold access to technologies needed for innovation.

Some observers, particularly academics, argue that recognition of a research exemption from patent law could avert the potential negative effects of upstream IP rights in biotechnology. Such an exemption, they hold, would be an important implement to prevent the "tragedy of the anti-commons"; without it, patents on gene fragments and other research tools could deter downstream research through the increased transaction costs of rearranging entitlement, the heterogeneous interests of owners, and cognitive biases among researchers (Heller and Eisenberg 1998; Eisenberg 2002).

These controversial arguments have motivated a number of major efforts to delve into the difficult and complex issues surrounding patents in the field of biotechnology. In 2002, the U.S. Federal Trade Commission conducted an exhaustive and comprehensive set of hearings featuring testimony from biotechnology industry representatives as well as academic witnesses. Though it recommended certain changes, the FTC concluded that the system, through a combination of incremental judicial interpretation and administrative action, had demonstrated remarkable self-correcting powers pertaining to trends and policies that could slow innovation (U.S. Federal Trade Commission 2002a; 2002b; 2003).

Others, both inside and outside the hearings, disagreed; and the sharp division of opinion prompted the National Academy of Sciences (NAS), through its Board on Science, Technology, and Economic Policy, to undertake a series of research projects. To exploit more fully the expertise of the U.S. research and private sector communities, the board established the Committee on Intellectual Property Rights in Genomic and Protein Research and Innovation. The board and its committees have produced several notable reports in the past several years, including *Patents in the Knowledge-Based Economy* (National Academy of Sciences 2003); *A Patent System for the 21st Century* (National Academy of Sciences 2004); and, in November 2005, a follow-on study, *Reaping the Benefits of Genomic and Proteomic*

Research: Intellectual Property Rights, Innovation and Public Health (National Academy of Sciences 2005).[2]

As a part of the research underpinning the 2003 report, the National Research Council (NRC) of the National Academies commissioned three independent academic experts (John Walsh, Ashish Arora, and Wesley Cohen) to analyze the situation and draw conclusions regarding the strength of an anticommons effect and the potential deleterious impact of patenting research inputs and tools. The basis for their subsequent report to the NRC was a series of extensive interviews with seventy IP attorneys, business managers, and scientists from ten pharmaceutical firms and fifteen biotech firms, as well as scientists and technology transfer officers from six universities, patent lawyers, and representatives from trade associations. The interviews focused on relations among universities, pharma companies, and biotech companies; the impact of patent policy on the behavior of firms; and recent changes in patenting and licensing activity (Walsh, Arora, and Cohen 2003a).

In their 2003 report, the trio acknowledged that the conditions for an anticommons problem might, indeed, exist in biotechnology. These included numerous patent claims on both inputs and final products that might increase transaction costs beyond the worth of the patents themselves; heterogeneity among the institutions holding patent rights—large pharmaceutical firms, small biotech firms, universities, large chemical firms, and IP holding companies—whose diverse goals and managerial experience would increase the difficulty of reaching agreements; and, finally, the uncertainty of the value of the rights, particularly in upstream research tools and discoveries, a situation that could readily produce asymmetric individual valuations that would cause a breakdown in negotiations (Heller and Eisenberg, 1998).

After extensive interviewing and subsequent analysis, however, the authors concluded that while "there [had] in fact been an increase in patents on the inputs to drug discovery . . . drug discovery [had] not been substantially impeded by these changes" (National Academy of Sciences 2003, 285).

Madey v. Duke University. In October 2002, just as Walsh, Arora, and Cohen were completing their report, and just over a year before the

2004 NAS report was published, the Federal Circuit Court of Appeals (a specialized court for patent litigation), ruled in a patent infringement suit against Duke University. The court decided that neither basic nor applied research is exempt from patent law, even when undertaken by academic scholars. Research represents the "business" of a university, one that brings increased funding and prestige (Fleischer-Black 2003; Maebius and Wegner 2002). Specifically, the court stated,

> Major research universities, such as Duke, often sanction and fund research projects with arguably no commercial application whatsoever. However, these projects unmistakably further the institution's legitimate business objectives, including educating and enlightening students and faculty participating in these projects. These projects also serve, for example, to increase the status of the institution and lure lucrative research grants, students and faculty . . . Regardless of whether a particular institution or entity is engaged in an endeavor for commercial gain, so long as the act is in furtherance of the alleged infringer's legitimate business and is not solely for amusement, to satisfy idle curiosity, or for strictly philosophical inquiry, the act does not qualify for the very narrow and strictly limited experimental use defense. Moreover, the profit or non-profit status of the user is not determinative.[3]

The *Madey* decision heightened concerns about the creation of an "anticommons" effect, especially in connection with research tools. In their report to the NRC, Walsh, Arora, and Cohen made a preliminary attempt to determine whether use of a de facto research exemption by universities had attracted numerous infringement suits from patent owners. At that time (before the effects of the *Madey* decision could be evaluated), the study found little support for apprehensions, concluding, "We . . . find little evidence that university research has been impeded" (National Academy of Sciences 2003a, 285).

The 2005 NAS Study. To accompany the 2005 NAS committee report, NAS officials commissioned a more extensive survey of 414 biomedical researchers in universities, government, and nonprofit institutions, conducted by two of the original researchers, Walsh and Cohen, and Charlene Cho. To a great degree, the results confirmed the 2003 findings. Only 1 percent of the random sample of academic researchers reported suffering a project delay of more than a month because of underlying patents, and none reported stopping a project due to the existence of third-party patents. Walsh, Cho, and Cohen (2005a) found that the main reason for the absence of patent impediments was that academic scientists—even after *Madey*—still paid little attention to patents in their fields (even among those 20-odd percent who had been notified by their institutions that such patents existed). In a synopsis of their research in *Science*, they explained, "Our research thus suggests that 'law on the books' need not be the same as 'law in action' if the law on the books contravenes a community's norms and interests . . . Our results suggest that infringement remains of only slight concern" (Walsh et al. 2005b, p. 2002). They concluded, "Our results offer little empirical basis for claims that restricted access to IP is currently impeding biomedical research."[4]

Recent Studies: 2006–07. Finally, since 2006 two additional studies have reinforced the analysis of the research commissioned by the NAS.[5] In 2007, David Adelman, from the University of Arizona, and Kathryn DeAngelis, from the law firm of Piper, Rudnick Gray, and Cary, published a detailed study of more than fifty-two thousand biotechnology patents granted in the United States between January 1990 and December 2004. They also presented findings from five subgroups. In the words of the two authors, their study described "the general trends in biotechnology patenting including patent counts, patent-ownership patterns, and the distribution of biotechnology patents across distinct areas of research and development." They concluded, "This analysis finds few tangible signs of patent thickets that define the anticommons" (Adelman and DeAngelis 2007, 6).

The reasons for this, the authors argued, were several. First,

> proponents of the anticommons theory presume, as they must, that the commons for biomedical science is strictly finite and congested. Yet a characteristic of biomedical science that stands out is its unbounded scope . . . The opportunities for

biotechnology consequently far exceed the capacities of the scientific community . . . It is this disparity between resources and
opportunities [that] makes biomedical science an unbounded
and uncongested resource. (24)

Second, there was widespread ownership, among diverse stakeholders,
and that characteristic showed no signs of changing, despite increasing
property rights:

> One of the [study's] most significant findings is the degree
> to which ownership of biotechnology patents is diffuse. Even
> the largest companies, on average, are granted fewer than
> thirty biotechnology patents per year, and the number of entities
> obtaining biotechnology patents has considerably increased
> over the fifteen years covered by the dataset. Interpreting these
> trends is necessarily impressionistic, but the lack of concen-
> trated control, rising number of patent applications, and
> the continuous record of new market entrants provide strong
> evidence that biotechnology patenting is not adversely affecting
> innovation. (3)

Thus, they concluded, proponents of the anticommons theory

> gloss over the conditions necessary for patent thickets to
> emerge, and their vivid metaphors obscure the complexities of
> interpreting patent-count data. In essence, they offer a one-
> dimensional model premised on a simple relationship existing
> between patent counts and transaction costs. (Adelman and
> DeAngelis 2007, 4)

For the second study, one member of the original NAS research team,
John Walsh, teamed up with three other analysts (Timothy Caulfield,
University of Alberta; Robert Cook-Deegan, Duke University; and F. Scott
Kieff, Washington University School of Law) to survey the current scholar-
ship and render conclusions based upon current conditions. Their argu-
ment, in a nutshell, was that policy recommendations for patent reform in

biotechnology have largely been driven by a small number of high-profile incidents and controversies (particularly the Myriad [BRAC-1-BRAC-2] case)[6]—and that these anecdotes do not accurately reflect the larger realities surrounding patenting in biotechnology:

> Our review of the lively policy debate and the limited empirical support for the claims that are driving that debate suggest that policymakers may be responding more to a high-profile anecdote or arguments with high face validity than they are to systematic data on the issues. (Caulfield et al. 2006, 1094)

Regarding the oft-stated fears of a developing anticommons logjam, the authors concluded:

> First, the effects predicted by the anticommons problem are not borne out by the available data. The effects are much less prevalent than would be expected if its hypothesized mechanisms were in fact operating. The data do show a large number of patents associated with genes . . . [One] study estimated that in the United States over 3,000 new DNA-related patents have been issued every year since 1998, and more than 40,000 such patents have been granted. But despite the large number of patents and the numerous, heterogeneous actors—including large pharmaceutical firms, biotech startups, universities and governments—studies that have examined the incidence of anticommons problems find them relatively uncommon.

Regarding the specific issue of patents blocking the use of upstream discoveries, they stated:

> The empirical research suggests that the fears of widespread anticommons effects that block the use of upstream discoveries have largely not materialized. The reasons for this are numerous and are often straightforward matters of basic economics. In addition to licensing being widely available, researchers make use of a variety [of] strategies to develop working solutions to

the problem of access, including inventing around, going off-shore, challenging questionable patents and using technology without a license . . . An anticommons or restricted access–type failure requires not that any one strategy be unavailable, but that the entire suite be simultaneously ineffective, which may explain why, empirically, such failures are much less common than was first posited.

The overall lesson from the recent reports is that, while the situation should be watched carefully in the future, at this time there is little urgency for sweeping, drastic legislative action.

Self-Correcting Remedies

A major factor in the lack of either a strong anticommons trend or the hindrance of innovation in biotech stems in part from the emergence of "working solutions," a variety of private strategies and public responses that have deterred large-scale adverse effects (Walsh, Arora, and Cohen 2003b). On the private side, evidence assembled by the researchers, and confirmed by events since the 2003 report was published, shows that cross-licensing, including the licensing of research tools, is relatively frequent. Most companies with patents on inputs or research tools have adopted liberal licensing policies, with relatively accessible fees and even some price discrimination in favor of university-based research. Other mitigating circumstances include the many research and commercial opportunities now available in biotechnology, which allow companies to move into research areas not plowed previously and, in a number of instances, give them the ability to "invent around" existing patents. When both the patent-holder and the prospective user know that inventing around is at least possible, there is strong inducement to settle on reasonable terms.

Farther out in the "informal" territory, university researchers often simply ignore upstream patents, in effect invoking by their action an ad hoc "research exemption." Many firms have been reluctant to enforce their patents against universities—particularly when the university is engaged in what seems to be purely noncommercial research—because of the prospect of low damage awards, the accompanying bad publicity, and the potential to jeopardize vital relationships with the research community.

Finally, there are more formal private mechanisms that can provide access to upstream patents: reach-through licensing agreements (RTLAs) and patent pools (U.S. Federal Trade Commission 2003; Walsh, Arora, and Cohen 2003a). RTLAs allow the original patent-holder to share the value of discoveries coming after the licensing of the patent (for a research tool, perhaps), usually through a royalty based on a percentage of sales of the final product. This can reduce the upfront costs for many small, as yet unprofitable, biotech companies, and promote risk-sharing between the original patent-holder and the biotech company. In the recent FTC hearings, two potential problems were identified with RTLAs: uncertain antitrust implications for these arrangements, and the danger of so-called "royalty stacking," a situation where so many royalties are attached to a final product that the equivalent of an anticommons results. This latter danger is potentially serious, but no evidence presented at the hearings indicated that the problem was widespread—nor did the FTC in its subsequent report see fit to recommend administrative or legislative changes in this area.

The creation of patent pools is a more formal arrangement than RTLAs. In this case, a number of patents are licensed together as a package, either through the leadership of one patent-holder or through the creation of a separate entity. Patent pools typically function by extending membership to industry firms that agree to assign or license individual patents, with members either giving each other royalty fee–licensing to all the patents or paying on a per-patent basis.

Some commentators have argued that patent pools are likely to be most successful when a number of horizontal competitors share similar values and practices and are forced by circumstances to engage in repeated transactions (Rai 2001). They concede that although the biotech industry has been composed of heterogeneous elements and interests (including large pharma firms, small biotech firms, universities, chemical companies, government agencies, and so forth), recent trends in the private sector leading to mergers and numerous tight alliances now make it more plausible that such pools might work. Conversely, critics have pointed out that significant anticompetitive issues can flow from patent pooling, namely, the exclusion of some firms from the pool, and grantback requirements that force pool members to grant licenses to each other for any technology developed from the original license.

Despite these possibilities, and after recent reviews of the issues, both the Justice Department and the USPTO have strongly encouraged this mechanism as a means of facilitating access to research tools (U.S. Patent and Trademark Office 2001a; 2001b; U.S. Federal Trade Commission 2003). They argue that any potential anticompetitive results can be dealt with adequately by existing antitrust laws and regulations.

Positive Institutional Responses

In the past several years, the U.S. Patent and Trademark Office, the judicial system (Court of Appeals for the Federal Circuit), and the National Institutes of Health all have revised administrative, judicial, and policy directives in response to some of the criticisms directed during the 1990s at the evolution of patent policy for biotechnology. This section will describe the most significant changes that have occurred.

USPTO. In 2001, the USPTO issued new utility examination guidelines for biotechnology, which in important respects tightened the requirements. The utility requirement in patent law stems from provisions that mandate disclosure by a patent applicant of "the manner and process of making and *using*" the invention (Section 112 of the Patent Act, 1952; italics added).[7] In most cases, the utility requirement has little effect, as no inventor would want to patent a useless invention. In chemistry and biology, however, the requirement plays a larger role, in that discovery in these fields typically involves identification of the product first and testing for its uses later. In 1966, the U.S. Supreme Court created a relatively strict utility standard by requiring a patent applicant to show that the invention has "specific benefits in currently available form."[8] The court stated that "a patent is not a hunting license. It is not a reward for the search, but compensation for its successful conclusion" (quoted in National Academy of Sciences 2005).

For the USPTO, the issue came up with regard to patent applications on expressed sequence tags (ESTs, or gene fragments) and how to handle those with unknown functions. The 2001 guidelines instruct patent examiners to reject claims for inventions that lack a "specific, substantial, and credible" utility, credibility being "assessed from the perspective of one of ordinary skill in the art." Thus, while the USPTO continues to hold that an "isolated and purified" genetic sequence is patentable, it must be accompanied by a written description of how it "can be used to produce a useful protein," or "serves as a marker for a disease gene," or "has a gene-regulating activity" (U.S. Patent and Trademark Office 2001a; 2001b).

Even critics of the present system have been somewhat mollified by the guidelines. One such critic, John Golden, has written,

> The PTO is right to head in this direction. Although the utility doctrine may not offer a realistic basis for substantial "roll back" of patentability, it can be deployed quickly to preserve the essential aspect of the existing balance between public and private interests . . . The new guidelines at least indicate that the PTO is committed to allowing the utility doctrine to realize its potential for substantial "bite." (Golden 2001, 34–35)

As this study was being completed, the USPTO demonstrated further determination to challenge broad patents—even if they had attained initial approval by the agency itself. In this case, the USPTO took preliminary steps to revoke three fundamental patents related to human embryonic stem cells that had been granted to the University of Wisconsin. Patent office examiners argued that the three patents appeared to be the same, or obvious variations of, cells (and cell activities) described in earlier scientific papers, or in patents issued to others. This move is only the first step, and the issues surrounding the stem cell patents are controversial—Wisconsin could still prevail in the end—but the action does signal a closer scrutiny for genomic patents by the USPTO in the future (Pollack 2007; Murray 2007).

Judicial Decisions. The courts are also drawing back from earlier expansive readings of patent scope and reach in biotechnology. Recent research has shown that in recent years the U.S. Court of Appeals for the Federal Circuit (CAFC) has gone from upholding the plaintiff in infringement suits in 60 percent of the cases to finding for the plaintiff in about 40 percent.

Many observers believe this is part of a trend away from siding with plaintiffs. Of recent cases, most often cited to underline this perception is *Regents of the University of California v. Eli Lilly and Co.*[9] The University of California had tried to argue that its patent on insulin, founded upon research with rats, also covered Eli Lilly's human-based bioengineering production process. The court found that because the university did not actually possess this claimed invention at the time of filing, it could not subsequently assert infringement, and, therefore, the claim was invalid.

In this case—and others—the CAFC adopted a strict application of "written description" and refused to allow the patent-holders wide-ranging claims to analogous sequences in other species (Rai 2001). In 2004, the court went further, in *University of Rochester v. G. D. Serle*, where it extended the strict application of the written description and demonstrated it could defeat claims seeking to reach through to future compounds that might be found through the use of protein structure information.[10] Though the case generated concurring and dissenting opinions, the ultimate resolution suggests that written descriptions, narrowly construed, will pose a significant obstacle to reach-through claims defined functionally, but not structurally (National Academy of Sciences 2005).

On a much broader front, the Supreme Court, in a unanimous decision on April 29, 2007—*KSR International v. Teleflex*—dramatically intervened in the debate over the current state of the U.S. patent system.[11] The issue was whether the USPTO had correctly applied the "novelty" standard in granting a patent for an automobile gas pedal that essentially combined new electronic mechanisms with an existing manual foot pedal. Writing for the entire court, Associate Justice Anthony Kennedy delivered a stinging rebuke to the CAFC and challenged the "constricted analysis" employed by that court. Kennedy specifically pointed to several errors and "fundamental misunderstandings" in its application of patent law.

KSR International v. Teleflex set forth several important principles. Most important, the Court mandated a "commonsense" approach to questions of novelty and nonobviousness. In the case before them, stated Justice Kennedy, the issue was whether "there existed at the time a known problem for which there was an obvious solution." The challenged Teleflex patent fit that description, in the Court's judgment. Kennedy's opinion went further and, in *obiter dicta* that presaged the future attitude of the Court, set out its views on the deleterious consequences of a relaxation of patent standards regarding nonobviousness and novelty: "Granting patent protection to advances that would occur in the ordinary course without real innovation retards progress," he wrote, and would deprive earlier inventions of "their value" and utility (as quoted in Pollack 2007; for other analyses of the Court's decision, see Zuniga 2007; *Economist* 2007; and Waldmeir 2007).

According to close observers of the Supreme Court, Chief Justice John J. Roberts has a strong personal interest in intellectual property issues. Over the past two years, the Court has accepted a half-dozen cases in this area, and more attention at this high level is expected in coming years (Sipress 2007, D1).

The NIH: Public Science and the Bayh-Dole Act. When large-scale public financing is introduced into the equation—as is the case with biotechnology and research funded by the National Institutes of Health—the calculations regarding costs and benefits of the patent system become quite a bit more complicated. Two events in the relatively short history of the biotechnology industry have assumed great importance in explaining the course of public innovation policy in this sector. The first, described earlier, was the decision of the courts to allow patenting of some basic research, specifically, the 1980 Supreme Court decision in *Diamond v. Chakrabarty*, holding that genetically engineered organisms were eligible for patent protection.[12] This decision, among other things, encouraged private firms to deepen their investment in basic research, and ultimately spawned thousands of small biotech firms that would try to build a business model upon the selling (licensing) of patented research results to other firms.

Linked to this far-reaching change was the passage by Congress of the Bayh-Dole Act of 1980 that codified as explicit public policy the encouragement of universities to seek patent rights for government-sponsored research. The reasoning behind the act was that society was not reaping adequate benefits from the huge federal investment in public research, and that the only way to increase societal payoff was to allow the direct beneficiaries of public research—mainly the universities—to patent their research and license these property rights to the private sector.

In retrospect, it can be seen that this legislation had an enormous impact on the goals and activities of U.S. research universities. As we saw above, when they seized the new opportunity for a potentially potent new stream of cash from patents (though never fully realized), universities plunged wholesale into entrepreneurialism. Patents to universities grew tenfold from 1980 to 1998 (from 380 to 3,151), with patents in biotech counting for about half of all patenting revenue. Further, more than half of the patents licensed to small biotech businesses were exclusive licenses (Eisenberg 2002). According to critics, these trends have produced in recent years a corrosive division in universities, torn as they are between their traditional defense of the great "scientific commons," the aim of which is to make the results of public research widely available, and their intoxicating new role as entrepreneurs, maximizing revenues from intellectual property rights. As economist Richard Nelson (2003) has argued, "In the era since Bayh-Dole, universities have become a major part of the problem, avidly defending their rights to patent their research results, and license as they choose." At the same time, ironically, it is the universities and their academic defenders who have led the charge for wide research exemptions. These issues continue to play out, as described in a recent article in *Nature Biotechnology* (Lawrence 2007).

National Academy of Sciences. It was against the background of the *Madey* decision and the shifting priorities of public and private interest groups that the National Academy of Sciences in 2004 evaluated recommendations for changes in public policy regarding patenting and licensing of university research, both privately and publicly financed. Though affirming its belief that the government should "consider providing some explicit protection from infringement liability" (National Academy of Sciences 2004, 110) for some research uses, the NAS report was striking for the candor with which it enumerated the immense difficulties and complexities of providing such protection without damaging the U.S. innovation system. It laid out four specific problems (110–11):

First, said the report, "not all activities that could be considered research deserve protection. Curiosity-driven inquiry that advances fundamental knowledge perhaps should not be subject to infringement liability, but R&D that is directed at commercializing the patented product should not be free to ignore intellectual property. Where to draw the line is far from obvious."

Second, "although much basic research is performed in universities, and companies tend to focus their [efforts] in applied research and development, there is no sharp division of labor, as the Federal Circuit observed in *Madey v. Duke University* . . . : 'Duke, . . . like other major research institutions of higher learning, is not shy in pursuing an aggressive patent licensing program from which it derives a not insubstantial revenue stream.'"

Third, "conversely, many corporate laboratories conduct fundamental research whose results are published in the peer-reviewed scientific literature. So if research meriting protection and research not meriting it cannot be clearly distinguished by who performs it or where it takes place, we are left with defining the difference and then trying to apply the definition on a case-by-case basis . . . This effort may have been feasible in an earlier era but before the distinctions between basic and applied research or between science and technology broke down."

And, fourth, "a further complication is that even within the realm of fundamental research there are activities that should not be shielded from liability. An example is the use of research tools whose development depends on the incentive provided by patent protection. How often this is the case is unclear, but . . . we should encourage . . . the observance of intellectual property to promote investment in the development of new and better research tools."[13]

Having acknowledged these difficulties, it is telling that the report failed to line up behind a single solution; rather, it retreated to analysis of the advantages and disadvantages of a series of proposals. In general, these proposals fell into two categories: those that suggested changes in the patent system, and those that more narrowly targeted language related to publicly financed research in the Bayh-Dole Act or to the powers of the NIH. Regarding the proposals to change the patent system, the NAS admitted that none was "problem-free" (National Academy of Sciences 2004, 115). Though it did recommend that Congress consider the options set forth, it assumed that no action would be taken, at least regarding changes in patent law.

As we have noted, the subsequent 2005 report requested a more extensive study of potential intellectual property impediments to the innovation process in biotechnology; and the independent researchers who performed that study, Walsh, Cho, and Cohen, again found "little empirical basis for claims that restricted access to IP is currently impeding biomedical research"

(2005b, p. 2002). Despite this, in 2005, the NAS committee dropped its previous carefully balanced stance and advanced specific recommendations for important changes in the regulation of genomic and proteomic research (National Academy of Sciences 2005).[14] The most far-reaching were, first, that Congress consider enacting a research exemption for research "on" patents if done to discover the validity of the patent; the features, properties, and inherent characteristics of the patent; novel methods of making or using the patented invention; or novel alternatives or substitutes for the patent; and, second, that by statute the standards for nonobviousness in the area of genomics be tightened and made more restrictive.

Regarding the second recommendation, as noted above, the recent Supreme Court decision in *KSR v. Teleflex* has likely started a process by which the standards of nonobviousness will be tightened through future judicial actions. As to the first, the next section will describe why we think *legislating* a specific research exemption for universities is unwise—or at least premature—at this time.

Bayh-Dole and the NIH. Both the NAS and outside commentators have focused their most intense interest on changes in the rules for publicly financed research governed by the Bayh-Dole Act and the NIH. The academy, doubtful that Congress would pass legislation in this area, has suggested administrative action that would allow the federal agencies to assume liability for patent infringement by investigators (universities) whose work it underwrites through grants, contracts, and cooperative agreements. This would be done through so-called "authorization and consent" provisions that make the patent a patent of the U.S. government. A variation of this proposal would limit "authorization and consent" to those situations where access to research-tool technologies cannot be resolved in the marketplace by licensing on reasonable terms.

Several academics have advanced somewhat different proposals. One suggestion, by Richard Nelson of Columbia University, is to amend the Bayh-Dole Act to grant immunity to universities from prosecution for using patented material in research if, first, those materials were not available on reasonable terms, and, second, if the university agreed not to patent anything that came out of the research, or, if it did so patent, to allow use on a nonexclusive basis (Nelson 2003).[15]

A second, similar proposal would also amend Bayh-Dole, giving the NIH clearer authority to oversee the patenting process with regard to the research it funds (Rai and Eisenberg 2003). Proponents of this change argue that the current language in the act drastically hampers NIH's ability to exert authority over patent licensing. They point out that under Bayh-Dole, NIH may restrict patenting only in *exceptional circumstances*, and that the statute provides for an elaborate procedure for appeal to the United States Claims Court. In addition, the agency must notify the commerce secretary, who has primary responsibility for the act and can overrule the NIH. Bayh-Dole also provides "march-in rights"—that is, the power to regulate commercial transactions related to products relying on the patented results of NIH-funded research—to the federal agencies, but only if the university is not taking steps to commercialize an invention or if the step is necessary to ensure public health or safety. There is no mandate that the move be "exceptional," but there is an elaborate administrative and judicial appeals process before final action.

Under the proposed amendments, the "exceptional" language would be deleted, giving the NIH broader and easier grounds for intervention in the licensing process. Further, invocation of "march-in rights" would be made less onerous, and the protracted delays in the administrative and judicial processes would be reduced.

The Way Forward for NIH: Better Safe than Sorry. Given the substantive complexity of the issues set forth here, the multiple interests in play, the huge contribution made by the biotechnology industry to health and to the U.S. economy, and the role of patents undergirding those contributions, caution should be the first principle in charting a future course. There are three reasons for this caution.

First, the empirical research undertaken by Walsh, Arora, and Cohen (2003a) and other independent researchers has produced little or no evidence of a growing "anticommons." No doubt, there are individual "horror stories" related to research tools or diagnostics (the 2005 National Academy of Sciences report exhaustively recounts several such incidents), but on the whole, the broad and deep U.S. science enterprise for biotechnology continues to move from strength to strength. Thus, self-correcting remedies and workout solutions seem to be inherent features of the current system, and to date have proved effective in overcoming barriers to innovation.

Second, regarding the NIH's power and authority, the agency has managed to get its way when it has believed that licensing proposals truly threatened competition and, ultimately, innovation, however cumbersome the administrative hurdles it has faced and however thin the legal bases for its actions. For instance, it negotiated with DuPont for more favorable terms of licensing of transgenic mice for the NIH and NIH-sponsored researchers. And it has pushed successfully for broader access to stem cells, as well as terms that precluded restrictions on publications via reach-through claims.

More recent actions reinforce our belief that it would be well to allow trends to play out before taking more drastic action. In 2004, the NIH finalized guidelines for licensing gene-related patents. The guidelines suggest that federally funded researchers should seek patents only when such inventions need "significant" private investment for commercialization, and that universities should adopt a general rule that patented inventions be licensed as widely as possible:

> Whenever possible, non-exclusive licensing should be pursued as a best practice. A non-exclusive licensing approach favors and facilitates making broad enabling technologies and research uses of inventions widely available and accessible to the scientific community. (U.S. Department of Health and Human Services, National Institutes of Health 2004, 18415)

The guidelines also make a clear distinction between rules for therapeutic and diagnostic applications, stating that

> patent claims to gene sequences could be licensed exclusively in a limited field of use . . . in therapeutic protocols. Independent of such exclusive consideration, the same intellectual property rights could be licensed non-exclusively for diagnostic testing or as a research probe to study gene expression under varying physiological conditions. (18415)

The main objective of the NIH document is to ensure widespread dissemination of research supported by the agency, a point stressed in the 2005 NAS report:

If an exclusive license is necessary to encourage research and development by the private sector . . . then the license should be tailored to promote rapid development of as many aspects of the technology as possible . . . If the licensee does not meet these milestones and/or progress toward commercialization is deemed inadequate, NIH recommends that the license be modified or terminated. Additionally, whenever possible, a licensing should include a provision allowing both the funding recipient and nonprofit institutions the right to use the licensed technology for research and educational purposes. (National Academy of Sciences 2005, 61)

In the current set of proposals to reform the patent system, little or no support for legislating some form of research exemption is apparent (except from some university organizations). In the unlikely event of a move in this direction, we would argue that one portion of the respective proposals (alluded to earlier in this section) by Rochelle Dreyfuss (2003) and Richard Nelson (2003) should be the starting point for negotiations—that is, if a university is given an exemption for a particular line of research, it should automatically forgo patenting rights for products that emerge from the research. In 2004 the NAS committee observed,

Dreyfuss's approach has the advantage of avoiding the need to characterize the invention or the manner of its use or to distinguish between exempt and nonexempt investigators by allowing researchers to self-identify. The government role would be limited to maintaining a registry of [patent authors'] waivers. (National Academy of Sciences 2004, 92)

On the plus side, the NAS committee noted that, "explicitly, the Dreyfuss proposal is intended to benefit university science and even in some degree to redirect faculty effort away from work with commercial applications or revenue-generating potential." On the other hand, the committee observed,

That runs counter to research universities' growing investment in technology transfer through patenting and licensing, encouragement of faculty to disclose inventions to central administrations, and aggressive pursuit of industry-sponsored research. Thus, one drawback of her proposal, acknowledged by Dreyfuss, is the friction likely to be generated or exacerbated between university administrators and researchers over when the waiver option should and should not be exercised. (93)

We would argue that this friction—or at least a fuller debate over the role of research universities and the consequences of commercial entrepreneurship on the scientific endeavor—would be a healthy, not a negative, outcome.

6

The Drive for Legislative Solutions

The previous chapter dealt with issues largely related to the biotechnology sector. But over the past two decades, hugely important changes have taken place simultaneously in the wider framework of patent policy in the United States. Some developments stemmed from challenges posed to the intellectual property regime by the emergence of new technologies, including biotechnology, and spinoffs from the continuing computer and telecommunications revolution in specific areas such as software and business methods. In addition, two significant institutional reforms produced unanticipated (and unintended) results that exerted a great impact on the U.S. patent system: the creation in 1982 of the U.S. Court of Appeals for the Federal Circuit specifically for patent litigation; and legislation in the early 1990s (the Omnibus Budget Reconciliation Act of 1990) that converted the U.S. Patent and Trademark Office into a service agency, no longer funded by federal appropriations but supported by fees collected from clients (patent applicants).[1]

A complete analysis of the role and impact of the CAFC is beyond the scope of this study. The rationale behind its creation stemmed from two quite separate sources. First, on a broader plain, during the late 1970s the fear grew that the United States was losing overall economic and technological competitiveness, and that a generally hostile attitude toward intellectual property was partly responsible (Scherer 2006; Landes and Posner 2003). Second, and more narrowly, major disparities and disagreements emerged in the handling of patent cases among the twelve federal circuit courts of appeal. Some courts—particularly those with a concentration of high-tech economic activity (California and Massachusetts)—were inclined to favor patent-holders; others (for instance, the circuit courts covering the Great Plains and the Midwest) were consistently more skeptical of patent claims. The result was increasing conflict and confusion and a "mad and undignified race" by applicants and alleged infringers to find a sympathetic appellate body (Jaffe and Lerner 2004, 100).

Certainly, the record of the CAFC since 1982 shows that it has worked to overcome the perceived flaws of the early 1980s—though in turn, as we shall chronicle, critics have maintained that it has created large new problems in the U.S. patent system. First, a "tilt" toward patent-holders (and against alleged infringers) is demonstrable: In the years prior to the establishment of the court in 1982, fewer than 30 percent of adjudicated patents were found to be valid; since 1982, this percentage has increased dramatically, varying from between 70 and 80 percent in some years to 50–60 percent in others. Over time, CAFC's new attitude toward patent-holders has been reflected in decisions of the federal district courts (where patent cases are initially argued): Whereas prior to CAFC, about 30 percent of patents were found to be valid and infringed by the district courts, afterward that proportion rose to about 55 percent. (For statistical reviews of the court's decisions, see Scherer 2006; Landes and Posner 2003; and Jaffe and Lerner 2004; 2006.)

Both supporters and critics of the new course set out by CAFC have identified several changes in the interpretation of patent law that have strengthened the hands of patent-holders. These include, first, a substantial increase by the court in the power and impact of remedies that can be exacted upon alleged infringers, through larger damages and relaxed rules for injunctive relief. Second, the court has expanded the number of topics eligible for patenting, the most notable being "business methods"; and, third, the court has limited the ability to challenge patent validity by loosening rules for "nonobviousness," thus allowing patent-holders greater leeway in asserting inventiveness (Jaffe and Lerner 2004).

Critics of the current system also point to the institutional changes at the USPTO as being partly responsible for the perceived decline in patent quality (National Academy of Sciences 2004). Although the fee system might well have provided adequate support for the USPTO, Congress has typically siphoned off a sizeable portion of the fees to fund unrelated items in the federal budget, rather than using them to improve the operation of the patent office.[2] As a result of this decade of underfunding, the USPTO is considerably understaffed, with each examiner having to review over one hundred patent applications a year. (By comparison, in 2001, the European Patent Office received 54 percent fewer applications but had nearly the same number of examiners.) One result has been ever increasing delays (pendancy of patents) over the past decade. In 2006, the average pendance was about 30 months, up 70 percent from the low of 18.2 months achieved back in 1991. For some types of patents (computer architecture, software, and information security), the USPTO can take an average of 44 months to reach a decision (U.S. Patent and Trademark Office 2006).

Moreover, the USPTO has faced continuing challenges in recruiting and retaining its examiners—by 2001, 55 percent of them had been at the office for two years or less. This is due mainly to the examiners' opportunities to leave the patent office and make considerably higher salaries in the private sector. During the years 2000–2006, the average annual attrition rate for the USPTO was 16 percent, compared to the average federal government rate of 6 percent (U.S. Patent and Trademark Office 2006; see also Jaffe and Lerner 2004; 2006).

A perverse incentive system also impairs patent quality. Under the current system, bonuses and promotions of USPTO examiners are based on productivity, which is measured by the number of patents that are reviewed and ultimately allowed or rejected. Because a patent can be appealed or revised if rejected initially, and therefore needs greater time to be processed, there is an incentive for examiners to "go easy" by approving more applications. It is estimated that examiners spend only sixteen to twenty hours on each patent. In addition, because each patent is unique, it is impossible for senior staff to review all the decisions of their junior colleagues, who often lack the experience and expertise to research the relevant prior art (that is, previous overlapping discoveries or patents).

Reflecting upon the unintended consequences of the institutional changes in both the legal and administrative agencies with oversight of the U.S. patent system, Adam Jaffe and Josh Lerner, two keen students of that system, concluded:

> It is now apparent that these seemingly mundane procedural changes, taken together, have resulted in the most profound changes in U.S. patent policy and practice since 1836. The new court of appeals has interpreted patent law to make it easier to get patents, easier to enforce patents against others, easier to get large financial awards from such enforcement, and harder for those

accused of infringing patents to challenge the patents' validity. At roughly the same time, the new orientation of the patent office has combined with the court's legal interpretations to make it much easier to get patents. However complex the origins and motivations of these two Congressional actions, it is clear that no one sat down and decided that what the U.S. economy needed was to transform patents into much more potent legal weapons, while simultaneously making them much easier to get. (2006, 2)

The profound changes described above have been accompanied over the past decade by an astounding burst of innovative activity in the United States, particularly in the areas of biotechnology, electronics, software, and telecommunications (Kortum and Lerner 1999; 2003). Combined with the relaxing of patent standards, this has had two further results of great relevance to this study: a substantial increase in the number and scope of patent applications and a dramatic rise in patent litigation.

First, consider the absolute increase during the past two decades both in the number of patent applications granted and in the number filed. Between 1930 and 1982, the number of patents granted rose at less than 1 percent a year, but from 1982 to 2002, the average annual growth was 5.6 percent, from 62,000 to 177,000 a year. In 2006, more than 400,000 patent applications were filed. These increases were accompanied by a seemingly inexorable climb in the incidence of patent litigation. The number of suits more than doubled between 1991 and 2001, rising from about 1,000 to 2,500 (though it has grown more slowly in the past several years). Not unexpectedly, litigation costs skyrocketed, too; a 2000 survey of intellectual property lawyers found that the cost of defending a large patent infringement suit (with more than $25 million at risk) was between $2.0 million and $4.5 million (American Intellectual Property Law Association 2001). A number of large technology companies spend over $100 million annually on patent litigation (Hedlund 2007).

Legislative Proposals

Discontent over the operation of the patent system has increased greatly since 2000, and, as we have described above, studies by highly respected institutions—the FTC and the NAS—have produced important recommendations for administrative, judicial, and legislation reforms. In the 109th Congress (2004–06), both the Senate and the House of Representatives considered bills that included proposals advanced by the NAS, FTC, interested stakeholders, and academics. Though hearings were held, and the bills went through several substantial revisions, in the end deep divisions among key stakeholders (particularly between the biotechnology/pharmaceutical groups and the high-tech software companies) blocked a legislative result.

In the 110th Congress, political control passed from the Republicans to the Democrats. There has, however, been no great partisan divide on many of the specific issues related to the patent system. At this point, although wide differences remain among interest groups, outside observers predict a higher likelihood than in previous years that patent reform legislation in some form will be passed. Following through on promises made in the last Congress to produce legislation this session, the chairmen of the relevant committees in the Senate and House introduced identical proposals on April 18, 2007, in the form of the Patent Reform Act of 2007 (S. 1145 and H.R. 1908). The bills were endorsed by the chairmen and ranking minority members of the two committees (Sen. Patrick Leahy, D.-Vt.; Sen. Orin Hatch, R.-Utah; Rep. Howard Berman, D.-Calif.; and Rep. Lamar Smith, R.-Tex.).[3]

The seemingly strong bipartisan support for the two bills obscures the reality that they represent merely the opening skirmish in battles to determine the final content and scope of a legislative package. Chairman Berman acknowledged this when he stated at an initial hearing in April 2007 that this is "not about a perfect bill"; it represents, he said, a chance for competing interests "to foster the policy discussion to yield the best result" (quoted in Munro and Noyes 2007, 56).

The Politics of Patent Reform

On April 18, *Washington Post* reporter Alan Sipress noted in a lead business section article that with "billions of dollars at stake, . . . Congressional initiatives to revise the patent system have drawn intense interest" and the "industries vying to sway the outcome have dramatically ramped up their campaigns, engaging some of Washington's most prominent lobbying firms since the start of the year" (Sipress 2007, D1). Other newspapers and magazines, similarly, are eagerly following the money trail and the tactics of "dueling industries . . . trying to elbow one another aside in their eagerness to steer legislators through the controversial process" (Munro and Noyes, 2007, 56).

Before describing the respective priorities and positions of key players in the legislative struggle, it is important to establish a fundamental underlying reality: While the patent system holds all industries substantively to the same rules, economic research, historical experience, and technological advances have demonstrated that the system affects and molds different industries in different ways (Levin et al. 1987; Schacht 2006). Thus, while rent-seeking is (as always) a driving force in the current debate, various industries are fundamentally reacting to the reform proposals based upon both the impact of the present system on their ability to compete and their projections of how new proposals will affect their future economic fortunes. With virtually no support for dividing the patent system along technological lines or by sector, the challenge to the present reform movement is to achieve a result that holds all participants to the same broad rules while retaining the flexibility to encompass both old and emerging technologies, and not tilt the system too far in favor of or against a particular industry.

Oversimplifying somewhat, we can say that in the present struggle, two polar industries—pharmaceuticals/biotechnology and software/information technology—best illustrate the dilemmas faced by the executive and Congress in putting together a viable and equitable legislative package.[4]

As we have discussed, the biotech industry perceives patents as the most critical element in protecting innovation. Innovation occurs generally in discrete steps, and a drug product usually embodies only one or two patents. Combined with the significant costs of R&D, the uncertainty of clinical trials, and the length of the regulatory process, patents are also judged to be critical to the biotech industry because of the relative ease of replicating the finished products, both in terms of costs and time. Not unexpectedly, both the biotechnology and the pharmaceutical industries oppose any legislative or administrative initiatives that, in their judgment, will weaken the patent protection afforded by the present system. In addition, biotech companies, which are often startups with little value beyond their embodied intellectual property, stress the overwhelming necessity of strong patents to attract indispensable venture capital.

The software industry starts from a very different position. In software (and related information technology industries), innovation is cumulative, with new products usually encompassing numerous patents—sometimes dozens or even hundreds of discrete advances. In addition, ownership of the embodied patents is divided among hundreds or even thousands of individuals or firms (today over twenty thousand software patents are granted each year in the United States; Bessen and Hunt 2004). Thus, the software and allied industries want above all to build increased flexibility into the system, unencumbered by injunctive delays and costly damages.

To some degree, the anecdotal fireworks over litigation—damages and injunctions—to date have obscured more fundamental reforms (such as a post-grant opposition system) that will be debated in the coming legislative process. This skewing is reflected in a recent *Washington Post* account, whose main theme comprises these issues:

> Large tech companies are more prone than many enterprises to trip over existing patents because the development of software is a fast-moving process that involves weaving together many small advances. So the computer industry seeks wider latitude to challenge patents while being protected against paying exorbitant damages, especially for unintended violations . . . But drug companies, which often spend years and billions of dollars converting just a few patents into highly profitable products, want strong rights to turn back challenges and to ensure that violators pay hefty damages . . . The drug and tech sectors, which rarely square off against each other in court, tend to play different roles in patents cases. Pharmaceutical companies are usually plaintiffs, while tech companies are more often defendants, and that difference explains their clashing views over the patent system. (Sipress 2007, D1)

The complexity of the political maneuvering surrounding these issues can be seen in the large number of coalitions, some of which go back to the earlier legislative drive in 2005, that have been formed to defend proposals in the current bills or push for changes. Pushing for the greatest changes in the present system are the software and allied information technology companies, combined in the Coalition for Patent Fairness. The coalition represents over seventy companies, including Apple, eBay, Intel, Cisco Systems, IBM, Hewlett-Packard, Microsoft, Sun, Symantec, the Business Soft Alliance, and several financial services firms, such as Visa. These companies are quite happy with the new bills as written in April, including provisions for tightened rules for injunctions and apportionment of damages for alleged infringement, an open-ended, post-grant opposition system to challenge patents without resorting to litigation, expanded scope of "prior rights" to give additional protection to trade secrets, and restriction of the venues for bringing patent infringement cases before federal district courts (U.S. Senate 2007b).

At the other end of the spectrum, opposing many of the provisions of the new legislation and pushing stronger patent protection, is, first, the alliance of biotechnology and pharmaceutical companies. They oppose a continuing "second window" for post-grant patent opposition, limitations on injunctive relief, and changes in the rules for apportioning damages in infringement cases, as well as delegation of larger rulemaking authority to the USPTO. In addition, the biotech/pharma companies want Congress to add provisions restricting the use of "inequitable conduct" against patent-holders and to repeal the "best mode" requirement, by which inventors must describe at the outset the best mode for utilizing their inventions (Biotechnology Industry Organization 2007; U.S. House of Representatives, 2007a).[5]

Allied with the biotech/pharma companies are the representatives of the U.S. research universities and the National Venture Capital Association (U.S. House of Representatives 2007c; Association of American Universities et al. 2007; Association of University Technology Managers 2006). Both also oppose an open-ended "second window," limitations on injunctive relief, and changes in the rules for damage apportionment. On some issues, however, both the universities and the venture capitalists differ with the biotech/pharma alliance. They oppose, for instance, the loosened rules for "prior art" defense,[6] arguing that this will allow large companies to protect trade secrets to the detriment of new patents for small companies and university researchers. They also both oppose restrictions on the ability of patent-holders to choose the venue for federal district court cases. Separately, the research universities—alone among the various interest groups—want the new legislation to include an "experimental research exemption," which would allow them to conduct limited research on patented products or processes.

Among the business coalitions, the very large Coalition for 21st Century Patent Reform is counted as more centrist and more flexible in the negotiations over the future course of S. 1045 and H.R. 1908, although its opening positions on key issues place it closer to the pharma/bio coalition than to the software/high tech Coalition for Patent Fairness. The quite diverse makeup of the 21st Century Coalition is no doubt behind its desire to seek compromise where possible; it includes companies that have experienced firsthand both sides of patent litigation—as patent-holders asserting rights and as defenders against infringement claims. The roster thus encompasses a number of sectors, from pharmaceutical firms to manufacturing and high-tech electronic firms, including Abbott Laboratories, Astra Zeneca, Bristol-Meyers, Pfizer, Caterpillar, Cargill, Dow Chemical, Exxon Mobil Corp., General Electric, Motorola, Texas Instruments, Procter and Gamble, Weyerhauser, and PepsiCo. Inc., among others.

So in addition to the consensus issues—a change to "first-to-file" facilitation of pre-grant opposition,[7] publishing of all pending patent applications within eighteen months, and permitting assignee filing—the 21st Century Coalition comes down on the side of the pharma/bio companies on most of the more controversial issues, such as the "second window" in post-grant opposition proceedings, apportionment of damages, injunctions, inequitable conduct, repeal of the "best mode" requirement, and limitations on venues where patent-holders can bring infringement actions. While this coalition does not deal with continuation proceedings explicitly, it does oppose granting "substantive" rulemaking authority to the USPTO (U.S. House of Representatives 2007b; Coalition for 21st Century Patent Reform 2007).

Finally, special note should be made of the influence of nonbusiness professional and governmental analyses and recommendations, particularly the positions taken by the committees put together by the National Research Council of the National Academy of Sciences and, to a lesser degree, the hearings and report of the FTC. Both the intellectual property section of the American Bar Association and the AIPLA also are providing important input to the legislative process (American Bar Association 2007; U.S. House of Representatives 2005).[8]

7

Evaluation and Recommendations

In this chapter, we shall evaluate the most significant individual reform proposals that are embodied in the current legislation or have been suggested by stakeholders or academic experts, according to the following criteria. To be effective, a proposal in our view must:

- **First, Do No Harm.** The most important lesson to be gleaned from earlier attempts to "reform" the patent system is the danger of unintended negative consequences from proposals advanced in good faith, such as those that resulted from the creation of the CAFC and the introduction of the fee system to support the USPTO. Headlong plunges into sweeping legislative changes may not be warranted since, as this study has chronicled, various institutions and decision-makers in the U.S. patent system have demonstrated a remarkable ability to identify flaws and implement self-correcting mechanisms and substantive policy changes utilizing existing authority (for example, the reforms instituted by the NIH and the USPTO). Thus, our first principle in evaluating new proposals for change is, "First, do no harm."[1]

- **Increase Information Flow through Bounded Adversarial Proceedings.** The history of the past decade makes clear that resources and institutional changes are needed to increase the flow of information throughout the patenting process, particularly for the examiners at the USPTO and for the federal courts, to expand their capacity to assess patent applications and claims of real innovation. One element of this drive is to introduce what we label "bounded adversarial opportunities" to the patent application process—"bounded," because we also are aware of the dangers of increased costs, protracted delay, and uncertainty that may result from new institutional proceedings.

- **Build upon Consensus among Major Stakeholders.** While consensus among stakeholders can be a sign of collusion against the public interest, long-standing and abiding substantive differences among key interest groups over the central issues of reform render this unlikely with regard to the patent system. Thus, pragmatically, we believe weight should be given to widespread support for changes by groups that are at odds on many other legislative proposals for patent reform. Examples (as we shall see below) include the change from a first-to-invent patent award to a first-to-file award; the publication of all patent applications within eighteen months of the date of filing; and the ability to assign patent rights to persons or institutions other than the actual inventor(s).

In setting forth here our views on the most significant provisions of proposed legislation, we shall first comment on proposals that have substantial support among major interest groups and have been vetted for some years. Second, we shall turn to what we consider the core elements of the reforms: the "bounded" adversarial institutional changes alluded to above. Finally, we shall discuss proposals that have evoked strong opposition from key stakeholder groups and which, in most cases, we think should not be enacted. We should add that, in making our individual assessments, we have attempted to look beyond each single action and envision the totality of a balanced package at the end of the legislative process. In the complex area of patent reform, with many contending interests and arguments, it is this final weighing of the patent-holder's need for certainty and stability against the need to preserve strong competition within the overall economy that presents the greatest challenge.[2]

Proposals Thoroughly Vetted, With Substantial Consensus

The proposals that have achieved a fairly wide consensus and have been thoroughly debated in recent years are the first-inventor-to-file system, the eighteen-month publication rule, and the grace-period provision. Each is discussed below.

Shift to a First-Inventor-to-File System. The "first-to-invent" system awarded a patent to the claimant who could prove that he or she was the first to "invent" a new product or process, while the first-inventor-to-file system grants priority to the inventor who first files with the patent office. The intellectual property section of the American Bar Association has stated that changing from the first-to-invent system to a first-inventor-to-file system "forms the core around which other proposed reforms to the patent system" will be assembled (American Bar Association 2007). This change enjoys widespread support among almost all major constituencies of the patent system and is considered central to the U.S. commitment to harmonize international patent rules. (In every other nation with a functioning patent system, priority of invention is dated from the earliest effective filing date.)

Though opponents (who are representative of individual small inventors) argue that a "race to the patent office" will hurt small inventors, most interested parties counter that the avoidance of lengthy and expensive legal proceedings (so-called interference proceedings) to discern just when the "Eureka moment" occurred takes precedence. The reform is particularly important for the biotechnology industry, which is characterized by multiple research efforts, often in the same or related fields, moving in parallel. Not surprisingly, research breakthroughs may occur at almost the same time, resulting in a high percentage of patent interference proceedings surrounding biotechnology patents. Determining who is the true inventor is often expensive, and the ensuing uncertainty and delay may last for months and or even years.

Furthermore, some observers have argued that small inventors and research institutions (including universities) are actually more nimble than large corporations. They note, in addition, the establishment by the USPTO of so-called "provisional applications" that allow individuals to obtain provisional property rights without undue expense or effort—and then later fulfill the more detailed obligations to obtain a patent (Holman 2006). Under this process, an inventor files a complete technical disclosure but then is allowed up to a year to refine the patent and develop commercial claims for the invention before submitting a formal application.

Finally, despite assertions of "independent inventor" representatives, recent research has demonstrated that the existing "first-to-invent" system has not generally worked to the advantage of small inventors; large companies and institutions have won a sizable percentage of interference disputes (Mossinghoff 2002; Lemley and Chien 2003).

Publication of All Pending Patent Applications within Eighteen Months. Current U.S. law provides an exception from the eighteen-month publication rule for patentees who plan to file only in the United States. All of the recent legislative proposals remove this exception—correctly, in our judgment. The eighteen-month rule completes a series of legislative and administrative reforms, begun in the 1990s, to reduce secrecy surrounding patent applications and bring U.S. practice more in line with foreign practices (Thomas and Schacht 2006). For our purposes, early public disclosure is basic to the goal of providing maximum feasible information regarding the patent application to USPTO decision-makers and interested outside interests, particularly for the purpose of pre-grant opposition.

Grace Period. Current U.S. law gives an inventor one year to decide whether patent protection is desirable. For instance, if an inventor first discloses the invention in a research article, he or she may file for a patent within a year from the publication date. This provision is particularly important for universities, where inventions often emerge from research not originally intended for patent protection. The grace period assures that the publication or other disclosure of an invention by an inventor who files an application within a year is not treated as prior art to that patent application, and that disclosures on the same subject by others during that period will also not be treated as disqualifying prior art. With the change to a "first-to-file" system, this rule is even more critical for universities and other research institutions.[3]

Assignment. Under current U.S. law, a patent application must be filed by an inventor, and the patent can be granted only to an individual. The rules govern those situations where the invention was developed by individuals in their capacity as employees. The proposed change would still provide that all inventors be named in the application, but would allow the inventors to "assign" their rights to their employers or other "parties of interest." This change was first put forth by patent commissions in the 1960s, and will bring U.S. law and practice in harmony with other nations.

"Bounded Opposition": Proposals to Introduce More Information into the System

Our recommendations to update the quality of U.S. patents while at the same time reducing uncertainty and controlling costs hinge upon carefully constructed administrative reforms in the patent application and review process. Specifically, we support greater opportunity to provide information and expertise to the USPTO before a patent is granted and a limited, but substantively wide-ranging, post-grant opposition. Our most important difference with the legislation just introduced in the 110th Congress regards what is called "second window" opposition, whereby at any time during the life of a patent an accused infringer can utilize the streamlined opposition system to challenge the patent.

Pre-Issuance Submissions. New reforms pertaining to the period after application for a patent and before the patent is issued are linked directly with proposals to publish all pending patent applications within eighteen months after they are first filed with the USPTO. In each case, the aim should be to provide the USPTO with all relevant information before it makes a final determination.

Under existing legislative authority, the USPTO has established a small window for members of the public to submit information regarding pending patent applications. In our judgment, however, the limitations are overly restrictive. The current rules provide that the submitted information can consist only of a patent or printed publication; nondocumentary evidence, such as sales figures or evidence relating to public use of the invention, may not be submitted. In addition, because Congress has mandated that no pre-grant opposition can occur without the consent of the patent-holder, the USPTO has stated it will not accept comments or explanations as to why the patent should not be granted. (The fiction here is that while it can accept outside information from anyone, it cannot utilize this information to establish a full opposition proceeding without the consent of the patent-holder.)

Provisions in 2007 legislation, which we support, would broaden the opportunity to oppose the pending patent before actual issuance, allowing any person to submit evidence of prior art. The submission would consist of documents relating to prior art, and supporting explanations in the form of "concise description(s) of the asserted relevance of each submitted document" (Thomas and Schacht 2007b, 31). The documentation must be submitted within whichever comes later: the date the USPTO issues a notice of allowance to the patent applicant; or six months after either the date of pre-grant publication of the application or the date of the rejection of any claim by the examiner. To assure that the process does not overly delay the application process, outside parties do not have the right to discovery or the right to argue their case before the examiners.

The hope and expectation for the opening up of the pre-grant submissions has been described by the IP section of the ABA:

> Where pre-issuance examination is complete, the issued patent can [we would substitute "may" here] avoid institution of a post-grant opposition. Where prior art has already been reviewed by the patent examiner—especially where the relevance has been fully investigated by the patent examiner—an opposer is unlikely to trigger the required threshold for instituting an opposition. (American Bar Association 2007, 3)

Post-Grant Opposition Proceedings. The provision of post-grant opposition and reexamination proceedings are not new ideas, either in the United States or abroad. European countries created such proceedings many years ago, and they appear to have operated successfully, introducing greater information into the examination system without at the same time gumming it up (Hall et al. 2003).

In the United States, Congress first created a rudimentary reexamination system in 1980. Initially, however, it did not allow outside parties to participate actively in the proceedings; in legal terms, they were conducted on an ex parte basis. Only in 1999, with the American Inventors Protection Act (AIPA),[4] did Congress expand the procedures to allow third-party (*inter partes*) participation. Still, the process has not been a success and has been only sparingly utilized (Thomas and Schacht 2007b). The reasons stem largely from Congress's understandable fear that challengers would game the system and frivolously attempt to block valid patents. Thus, it erred in the direction of constructing safeguards for the patent-holder that made utilizing the process very unattractive to challengers. Most important was the stipulation that in any later litigation over the patent, challengers were legally barred from presenting arguments that they made *or could have made* during the reexamination proceeding (even if the arguments were never considered by the USPTO). Furthermore, as with pre-issuance proceedings, evidence could only be in the form of other patents or printed publications. Finally, the 1999 AIPA allowed the patent-holder to appeal a negative determination to the courts, but did not allow the same right to the challenging party. The result is that only a small fraction of new patents have been challenged under this new third-party process (in 2002, only 25 of the 190,000; see Jaffe and Lerner 2006).

Legislative Proposals in the 110th Congress

The legislative proposals submitted in the 110th Congress demonstrate a strong consensus for a limited and carefully crafted post-grant opposition that would increase the efficiency of the patent system by precluding, in most cases, lengthy and costly court proceedings. There are, however, substantial disagreements about the details, scope, and tenure of such a system. What would the most equitable and efficient post-grant opposition system look like, and how would it differ from that provided by the current version of the legislation?

An Optimum System. In the most equitable and efficient post-grant opposition system, any interested party should be empowered to challenge the patent and bring forward all information pertaining to a wide range of concerns, including allegations of double patenting, challenges on grounds of novelty, nonobviousness, and other statutory provisions. (It should be noted that the 1999 AIPA already requires the USPTO to find "substantial" new evidence, or it will not initiate proceedings.) If the patent survives the challenge, parties in any subsequent litigation should only be barred from making arguments that were specifically advanced and rejected by the USPTO in the post-grant opposition process. (This would remove the strong disincentive attached to the subsequent barring of arguments "that could have been made" in the earlier proceeding.) To safeguard against delay, avoid "fishing expeditions," and prevent the parties from gaming the process, challengers should be required to submit all of their evidence at the outset, and patent-holders all of their rebuttal evidence in a single batch of documents. The USPTO would have discretion to establish rules for limited discovery. Both parties should have the right to appeal the USPTO's decision to the federal courts. (For more detailed descriptions of the procedural safeguards, see Thomas and Schacht 2007b and American Bar Association 2007.)

Finally, the post-grant proceedings should be given over to a "second set of eyes," not those of the original examiner. Proposals have ranged from having a single second examiner to appointing a panel of three administrative patent judges (the route taken in legislation in the 109th and 110th Congresses). The 2007 legislation establishes a patent trial and appeal board to oversee the post-grant proceedings.

Disagreement with Proposed New Legislation. In the area of post-grant opposition, we have three significant disagreements with S. 1145 as introduced. All relate to the ultimate balance between the rights of the patent-holder and those of challengers.

First, we would resurrect provisions in legislation in the 109th Congress regarding attorneys' fees and fees for triggering the post-grant opposition process. We think that a nominal fee ($50,000) should be required of the challenger, and that if the patent is successfully challenged the patent-holder should be required to reimburse the challenger for this fee, plus assume responsibility for all legal costs. Conversely, if the challenge fails, liability for the fee and legal costs should shift to the challenger. For both sides, this relatively modest change would provide at least some disincentive for frivolous action. For the challenger who just wants to probe and gum up the process, the possibility of both legal costs and losing the upfront fee provides some deterrent; a patent-holder who knows that the patent is weak or probably indefensible is also deterred by the prospect of paying the costs of merely delaying an adverse ruling.[5]

Second, of greater importance, we believe that the decision to open up a "second window" for post-grant opposition challenges, as contemplated in early versions of S. 1145 and H.R. 1908, will have substantial negative consequences—certainly for biotech companies, but for small R&D companies in other fields as well. It will tilt the balance too far, in our judgment, toward challengers.[6] As this study has noted, holders of biotechnology patents argue—correctly, in our judgment—that their ability to obtain research and development support for new products depends on investor confidence based upon strong patent rights. Allowing a second window to open at any time during the life of a patent would introduce a debilitating level of uncertainty throughout the entire patenting process. Further, it would likely embolden alleged infringers to challenge patent validity more often, rather than settling to avoid the large costs of utilizing the alternative route of patent litigation. In biotechnology, one of the chief beneficiaries of a lifetime second window would be generic drug manufacturers, who would have particularly strong incentives to challenge the validity of a particular drug before the end of the legal patent term.

It might well make sense to extend the time for a "first-window" examination by some months to give any challenger more opportunity to assemble a legal team and the data to support an opposition case (the new legislation allows a window of twelve months); but once this period for reexamination is ended, that should be it. Redress by the courts should then be the only recourse for a challenger or alleged infringer.

Though we admit it is a close call, our third dissent relates to changes in the new legislative proposals regarding the standard of review during the post-grant examination proceedings. Under existing law and practice, a patent, once granted by the patent office, is entitled to a "presumption of validity." This means that anyone challenging the patent must prove by "clear and convincing evidence" that it is invalid. The "clear and convincing evidence" standard is higher than the "preponderance of evidence" standard a patent-holder must meet to win claims of infringement. By removing the presumption-of-validity legal doctrine during the post-grant examination proceedings, S. 1145 and H.R. 1908 substitute the lower "preponderance of evidence" standard for the higher bar of "clear and convincing evidence."

There are two important reasons to argue against this lower standard. First, as a general principle of administrative law, issues that have been examined before a competent administrative body should be presumed to have been decided correctly. In an effort to tilt the playing field toward the challenger, those who constructed the new legislation have ignored long-standing procedures and doctrines which hold that after an open, competent administrative process, the regulatory body should be afforded a great degree of deference. Second, while it was possible previously to argue that the existing procedures did not allow for a fair and balanced assessment of the complicated issues presented to the USPTO and its examiners, the changes proposed above—more detailed pre-grant submissions and more resources for the patent examiners—render such a judgment much less credible.

Assuming only a "first window," however, several arguments have been put forward for the lower standard of proof. Many legal authorities (who oppose a "second window") are willing to accept the lower standard, arguing that the proceeding should actually be considered an extension of the patent examination, providing a brief period to clear up uncertainty or rectify mistakes. They also contend that the lower standard of proof would induce challengers to come forward early, thus weeding out bad patents more expeditiously and decisively confirming valid ones. While these are legitimate points, on balance we would still support the higher standard.

In any case, we strongly hold that if any form of "second window" is allowed, the higher "clear and convincing evidence" bar should be upheld. As Jaffe and Lerner have argued,

> There is . . . an important reason to maintain the presumption of validity. Remember that the fundamental purpose of the patent system is to give inventors a basis for expecting that they will have an opportunity to recover investments that they make in developing and commercializing their invention. When a start-up firm goes out to raise money for this purpose, it is important that the patent or patents that are claimed as the basis for protecting the firm's technology have the presumption of validity. If, instead, the validity issue were reduced to a legal coin flip, it would greatly increase uncertainty. Uncertainty is the enemy of investment, so patents of uncertain validity would be much less effective in providing a base for development of innovations. For this reason, eliminating the presumption of validity is a potentially dangerous

change in terms of its long-run consequences for the innovation process. (Jaffe and Lerner 2006, 22)

Subjective Elements of Patent Litigation. Several elements of U.S. patent litigation call for a judgment of a party's state of mind, either as a patent applicant or as an alleged infringer. Like the National Academy of Sciences, we believe that these peculiar elements should be removed or, at a minimum, modified.

Best Mode. All of the legislative proposals introduced in the last two Congresses would remove the requirement in U.S. patent law that the inventor set forth the best mode contemplated by the inventor of carrying out his invention. Failure to disclose this best mode has been a ground for invalidating the patent. While some observers have claimed that demanding the disclosure of best mode allows later inventors to compete with the patentee on a more equal basis, this provision has been severely criticized by recent commissions, academic studies, and all of the major stakeholders in the patent system (Schacht and Thomas 2006). Most now argue that, over the life of the patent, what constitutes best mode may well change as technology evolves and knowledge increases. Further, as now applied, the doctrine places the courts in the difficult, if not impossible, position of discerning the "subjective mind" of the inventor at the time he or she created the invention. Finally, as proponents of the change have pointed out, the patent law will still contain a strict requirement for a written description setting out exactly what was invented, and an enabling requirement that mandates enough detail for imitators to replicate the invention using routine methods and procedures.

Inequitable Conduct. Under existing U.S. patent law, an applicant is obliged to maintain candor and truthfulness in presenting written and oral material to the USPTO. Patent law penalizes those who violate this obligation under the doctrine of "inequitable conduct." If USPTO reaches such a finding, the patent is declared unenforceable. In recent years, accused patent-infringers have begun routinely raising this defense against patent-holders, so much so that the Federal Circuit has argued that "the habit of charging inequitable conduct in almost every major patent case has become an absolute plague" (Schacht and Thomas 2005, 29). Many stakeholders in the patent system have called for reform in this area.

The Senate patent reform legislation in the 109th Congress attempted to make two important changes in this area. First, it codified the "duty of candor" and gave a clear legal mandate to the USPTO to penalize conduct falling short of it. Second, however, it attempted to limit the circumstances in which such a defense can be raised by alleged infringers by barring such a pleading unless the court has found that at least one patent claim is invalid, and that the invalidated claim would not have been issued "but for" misconduct. Some (the IP section of the ABA, among others) have argued that the earlier Senate bill was flawed, and there is a good deal of discussion on how to achieve the ends described above. We leave it to the legal experts to work out, but in our judgment the rationale behind the proposed changes was sound. The "plague" alluded to by the Federal Circuit is real, and Congress should not ignore it as it moves toward major patent reform. (For more details on the adverse consequences of the current situation, see American Bar Association 2007; National Academy of Sciences 2004; and U.S. House of Representatives 2007b.)

Willful Infringement. Under current patent law, courts may increase the damages for infringement up to three times the amount assessed when a finding of "willful" infringement is reached. The circumstances that sustain a judgment of willful infringement are highly case-dependent and often force courts to attempt, once again, to assess subjectively the "state of mind" of the accused. In addition, a number of observers have argued that the lack of clarity in the criteria for determining willful infringement has the perverse effect of inducing more litigation by encouraging inventors to avoid searching out existing patents or proof of prior art for fear of incurring multiple damages. Fear of increased liability has also discouraged some firms from challenging dubious patents.

The 2007 legislation attempts to remedy the alleged faults regarding willful infringement. First, henceforth, determination of willful damages by a court can only occur after the patent has been found infringed, enforceable, and valid. Further, in order to prove willful infringement, a patentee is required to submit evidence that the infringer received written notice of infringement; that the infringer intentionally copied knowledge from the patentee; and that the infringer continued to infringe after the court ruling. In addition, infringement cannot be found if the alleged infringer can demonstrate that he or she possessed a good faith belief that infringement was not occurring. While no doubt there will be extended negotiations over the exact language, the goal is valid and the current legislation an advance toward greater clarity and predictability in the patent system.

Let the Process Work—or the Courts Decide

There are two areas—apportionment of damages and preliminary injunctions—where, in our judgment, Congress should stay its hand and allow developing case law to proceed.

Apportionment of Damages. Patent law provides that all patent-holders are entitled to damages adequate to redress the losses from patent infringement. The minimum level of damages that can be awarded to the inventor is a "reasonable royalty" for the use made by the infringer of the invention. The determination of the proper level of damages is highly fact-dependent. For instance, an infringed product or process may contain a great number of additional elements beyond the patented invention at issue. Consider a patent at issue that is related to a single component in any audio speaker, while the product being defended consists of an entire stereo system. Under current law and practice, the courts may apply a so-called "entire market value rule," which allows damages to be based upon the entire product when the quality of the audio speaker component was the central basis for consumer demand. Alternatively, the court may decide that demand was based on many other factors besides the audio component. In this case, it may apply principles of "apportionment" to calculate the amount of damages from the infringement.

Many think that in recent years the courts have systematically overcompensated patent-owners, and a number of horror stories are cited, particularly in the software and high-tech electronic industry. The issue is complicated by the many instances in which the decisions are left up to juries, which are prone to go off the deep end. In the oft-cited

Alcatel-Lucent v. Microsoft suit, for example, a jury ordered Microsoft to pay $1.52 billion for infringing two patents for the MP3 technology that is used to play digital music on computers, portable players, and other mobile devices.[7] The jury assessed the damages on Microsoft's worldwide sales rather than just its U.S. sales and based them on the value of nearly all computers with Microsoft Windows operating systems rather than on the far lower value of the patented MP3 technology (Sipress 2007).

Beyond the problem of jury ignorance or willfulness is the reality, once again, that products in the high-tech electronic and software industries most often consist of multiple patents, no one of which governs the competitive edge of the product itself—and that products in the biotech/pharma industries are usually made up of only a few patents, with the combination of patents often what constitutes real invention and the market edge.

The 2007 patent reform legislation, as introduced, tilts the playing field toward the high-tech electronic and software industries. Without getting into the legal weeds, we can say that, in effect, the legislation pushed the courts toward recognizing only the value of the individual patent components of a product, limiting their ability to assess the value of combinations or of the entire value of the product.

For over three decades, the governing case law in this area has been a federal district court opinion, *Georgia-Pacific Corporation v. United States Plywood Corporation*, which listed some fifteen factors pertinent in determining reasonable royalty damages.[8] Most observers believe the listed factors provide a sensible and equitable set of guidelines. Several of the coalitions and professional groups involved in the current legislative process have offered language that would codify *Georgia-Pacific* or clarify the intended balance between "apportionment" situations and cases in which the "entire market value" rule should be applied. Without endorsing specific language, our judgment is that this balancing goal is the correct way toward a means to satisfy the disparate interests of contending parties. Thus, if a legislative solution is finally deemed necessary, we would agree with the position of the IP section of the ABA:

> The Section supports the enactment of legislation permitting apportionment of reasonable royalty damages in a manner that protects infringers against unjustified damages awards yet

ensures that the value of the patented invention appropriated
into the infringing product or process is fairly recognized in any
damages award. The Section opposes legislation providing that
a determination of a reasonable royalty in the case of a combination patent shall be based only upon such portion of the total
value of the combination apparatus as is attributable to the
patentee's specific contribution over the prior art. (American Bar
Association 2007, 54, italics added; see also U.S. House of Representatives 2007b and Thomas and Schacht 2007a).[9]

Preliminary Injunctions. The 2007 patent reform legislation, as introduced, contained no changes in rules regarding injunctions after patent
infringement has been found. In 2005, however, some legislative drafts did
include provisions weakening the presumption in favor of injunctions in
patent infringement cases. And among interested parties, agitation remains
for changes that reflect the language of the 2005 drafts.

Much has changed in this area, however, since 2005, not least as the
result of a 2006 Supreme Court decision (*eBay Inc. v. MercExchange, L.L.C.*)
that may have redrawn the parameters for granting injunctive relief in
patent infringement cases.[10] Briefly, a jury returned a verdict of infringement against eBay, but the U.S. District Court for the Eastern District of
Virginia refused to issue an injunction, giving several reasons for its denial
including, first, that monetary damages would be an adequate remedy in
this case, and, second, that the public interest would not necessarily be
served by an injunction because MercExchange did not utilize its patents,
but existed "merely to license its patented technology to others" (Yeh 2007,
7). On appeal, the CAFC affirmed the infringement, but ruled that
MercExchange was entitled to an injunction, arguing, "Because the 'right to
exclude recognized in a patent is but the essence of the concept of property,'
the general rule is that a permanent injunction will issue once infringement
and validity have been adjudged" (as quoted in Yeh 2007, 7)

In May 2006, the Supreme Court unanimously vacated the CAFC judgment and remanded the case back to the district court. The court took no
position on whether an injunction was justified in this particular case, but
it ruled that the principles of equity governing injunctive relief "apply with

equal force to disputes arising under the Patent Act" (as quoted in Yeh 2007, 7–8), thus knocking down the notion that patent disputes were subject to a different set of rules on injunctions than other areas of law. For this study, two points in the court's decision are important: First, the court stated that the district court had been in error when it held that injunctive relief was categorically unavailable in instances where patent-holders only license their patents rather than work them themselves (a particularly important decision for universities who hold patents only for licensing); and second, as noted above, it slapped down the view of the CAFC that patent cases demanded a different standard for injunctive relief.

For patent-holders—such as biotech and pharmaceutical companies—*obiter dicta* from Chief Justice John Roberts (with Justices Ruth Bader Ginsberg and Antonin Scalia concurring) portended continuing sympathy for injunctive relief. Roberts predicted that injunctive relief would remain the usual remedy for patent infringement, consistent with a long "tradition of equity practice" (Yeh 2007, 8). It is hard to gauge exactly where the Supreme Court will go after this decision, but it—and lower courts—seem likely to move toward some kind of rule of reason, rooted in the facts of individual cases.

Thus, at this point there is much less reason for congressional intervention.

Substantive Rulemaking Authority and Continuation Proceedings. Two separate, but linked, proposals for change call for comment: new substantive rulemaking authority for the USPTO and reforms proposed by the USPTO for continuation proceedings. The new rulemaking authority is included in the 2007 bills; changes in continuation regulations were proposed originally in 2005 legislation and then subsequently by the USPTO through agency rulemaking.

Substantive Rulemaking Authority. The joint 2007 House and Senate bills, as originally introduced, would grant the USPTO substantive rulemaking authority in addition to its existing authority to make rules governing the operations of its office, a measure that would greatly increase the power and independence of the USPTO. Rules and determinations of the office would then have the "force and effect of law" and would be entitled to the deference granted to agencies under the famous *Chevron* decision.[11] This higher level of deference would, in turn, mean that a court would uphold a USPTO decision unless it appeared "unreasonable"—a high standard for reversal. As an example, rather than issuing guidelines for interpretation of utility or obviousness under existing statutes, the office would draft substantive rules which would have the effect of law unless the courts did not find them reasonable because of some underlying flaw. Beyond such examples are currently a large number of contentious issues surrounding the patent application, approval, and challenge processes that might be handled through this broad new grant of authority to the office.

Proponents of substantive rulemaking power for the USPTO argue that, within the federal government, this grant of authority is by no means novel. Other agencies with similar levels of expertise—the FDA, the FTC, and the U.S. Securities and Exchange Commission (SEC)—have been granted the power to use formal rulemaking to interpret general legislative mandates for a long time. Furthermore, those in favor of the grant hold that the additional powers will result in more sophisticated and flexible rules, as the office will be able to react with greater depth of expertise to changing technological conditions. Finally, they claim that the authority will work to reduce the inefficiencies and long delays of the U.S. patenting process.

Opponents, however, respond with a sharply different perspective on the potential consequences of what they consider a sweeping and ill-considered change. For our purposes, the arguments of the IP section of the ABA will suffice. First, the ABA points out that

> the patent laws reflect a delicate balance of competing policies. Over the last 200 years, the patent laws have been amended as a result of considerable public debate and discussion by elected officials in Congress . . . Providing substantive rule-making authority to appointed PTO officials removes this debate from elected officials . . . The Bicameral Bill [2007 legislation] simply grants too much authority to the PTO to promulgate rules that are best debated in Congress. (American Bar Association 2007, 63)

Second, the courts, not the PTO, are best able to handle ambiguous areas of patent law: "The Federal Circuit," says the ABA, "has developed expertise on the patent laws and is well-equipped to review PTO action."

And, third, the process is incremental and thereby results in great security and certainty: "Perhaps more significantly, the development of case law is incremental so that the law tends to change slowly, which provides an important level of certainty and predictability [to] users of the patent system" (American Bar Association 2007, 63–64).

At this point, it is our judgment that because of the relative suddenness and late arrival of what is admitted by all sides to be a considerable break with the past—combined with sharp differences over the consequences—substantive rulemaking authority should be removed from current legislative proposals and further study and debate fostered before a final decision is made. As with the question of continuation, which we will take up next, we recommend that the National Academy of Sciences be asked to conduct a thorough study of the issues surrounding substantive rulemaking authority for the USPTO and report back to Congress, stakeholders, and the general public within a certain time.

Continuation Proceedings. The continuation procedure permits inventors to restart the patent examination process while retaining the filing date of the original application that discloses the same invention. Inventors use continuations to revise their claims based upon technological changes that have an impact on their proposed patent, and to respond to examiners' findings and comments. There are three types of continuations: the "continuous application" (CAP), which discloses the identical invention claimed in the prior or "parent" invention and cannot include anything that would constitute new matter; the "continuation-in-part" (CIP), which contains at least a substantial portion of the original application but includes additional matter; and a "divisional" application, which is filed when the original application contains more than one independent invention. For divisional applications, the USPTO permits applicants to elect one of the disclosed inventions for continued examination, while the others can be withdrawn or amended and pursued in new applications called "divisions."

According to the USPTO, approximately 36 percent of all applications (115,000 of 317,000) in fiscal 2005—a representative year—were continuing applications. In January 2006, arguing that it must reduce workload and the substantial backlog that has accumulated and that continuing applications constitute an insupportable burden on the system, the USPTO proposed new rules placing much greater obligations with regard to defense and disclosure on continuing applications beyond the first one:

> In particular, the proposed rules require that any second or subsequent continuing application show to the satisfaction of the Director that the amendment, argument, or evidence could not have been submitted during the prosecution of the initial application or the first continuing application. (71 FR 48, January 3, 2006)

In addition, the patent office stipulated that if an applicant did not designate ten claims, then only independent claims would be examined initially (ibid.).

As noted, the rationale for these new restrictions was limited to the necessity to reduce the USPTO's workload and application backlog. No attempt was made to broaden the analysis to include implications for patent quality or strength, or for disparate effects across industries.

We do not propose to describe in detail the arguments, pro and con, regarding the continuation procedures in the United States. We can say, briefly, however, that critics of the current system point to a number of flaws and potentially anticompetitive results. First, it is argued that continuations breed delay and uncertainty in the patent process: The mean time for prosecution for a patent with one continuation is double that of patents with no continuation (and some few multiple continuations can last a decade; Lemley and Moore 2004). Second, it is claimed that continuation abets, even invites, strategic behavior—specifically, "submarine" patents that are issued after extended periods of examination and revision through continuation. Patent-holders can observe technological and market developments and then amend the patent through continuation to thwart competitors. Third, and more directly related to the biotech and pharmaceutical industries, are the allegations related to "evergreening," or instances in which patent-holders use continuation to challenge generic competitors by obtaining multiple variations of the same patent through sequential small changes that extend protection against competition from generic replicas (Lemley and Moore 2004).

Operating from a very different universe, biotechnology industry groups and their patent attorneys point out that, for this industry, continuation is essential because of the complicated and extended duration of the innovation and commercialization process. They note that many biotech companies start as spinoffs from academic discoveries and depend heavily on attracting investors for high-risk ventures. Thus, the pressure is relentless to patent early in the discovery process and then to protect the initial concepts and their subsequent embodiments through the continuation process. (The costly and time-consuming alternative would be to file multiple stand-alone applications.) As the Biotechnology Industry Organization, the industry trade association, has noted:

> This competitive pressure drives smaller biotechnology companies to file patent applications on inventions early in the development stage so that they may obtain that first patent to generate investor interest and to meet milestone markers established by investors. Consequently, biotechnology companies file patent applications years before a product or technology has been fully developed or commercialized. During this time, they may agree to initial narrow patents and continue to perform "proof of concept" experiments to further support their initial discovery. With the initial patent in hand, patent owners can point to other pending applications [continuations] that may be broader and more comprehensive to secure further investor interest. (Biotechnology Industry Organization 2006, 4)

In addition, most biotech patents are for medicines for human use and must be tested in lengthy clinical trials. Because the patent office requires correlative evidence for patent claims for human use, experiments to prove efficacy and safety often result in additional continuation claims.[12]

Some recent administrative changes and some proposed legislative changes will reduce, if not remove, some of the negative consequences of the U.S. continuation procedures. For instance, in 1995 Congress changed the patent term from seventeen years from patent issue to twenty years from the filing of the application. Then, in 1999, Congress required that for most patents, publication by the USPTO must come after eighteen months subsequent to the initial *filing* of the application (and the pending legislation would expand this publication requirement to all patents). Changing the patent term to begin with application has reduced the incentives to delay prosecution of the patent, and the eighteen-month publication rule will reduce the likelihood that secret submarine patents can surface years later.

Among the other alternatives to the USPTO-proposed rules that have been advanced are the following:

- Achieving greater flexibility in the system by introducing a deferral/acceleration system that is utilized in other countries. Under such a system, an applicant would be able to file inexpensively and then decide later whether to request full examination, allowing conditions to make this determination. Conversely, an applicant, under prescribed circumstances, could request an accelerated examination, through a so-called "rocket docket" process. These changes would allow the patent office to manage its resources and scheduling better.

- Limiting the number of years an application could spend in prosecution (possibly to eight years, for example; see Lemley and Moore 2004).

- Establishing a limit on the number of continuation applications, but raising that limit well above that proposed by the USPTO, possibly to as high as ten. Under such a proposal, the USPTO could be granted the authority to deal with the original application and amendments on a case-by-case basis, thus allowing it to shut down the process if there is evidence of strategic behavior.

We do not propose to support any particular remedy, but we are convinced that on this issue, as with the proposed grant of substantive rulemaking for the USPTO in pending legislation, more study and analysis are needed before a definitive judgment can be made. Thus, once again, we recommend that the USPTO seek the advice and recommendations of the National Academy of Sciences before publishing a final rule. (Under present circumstances, it is inevitable that the USPTO rulemaking will be challenged in the courts, so seeking broader support and advice before attempting to enforce a final determination may save time and produce a better decision in the end.)

A forthcoming paper by researchers with solid backgrounds in this area reinforces this recommendation. In their empirical study of why and under what conditions U.S. firms utilize the continuation process, Hegde, Mowery, and Graham first point out that "the debate over the use, abuse, benefits, and costs of the continuation procedure has been conducted in an evidentiary vacuum" (2007, 3). They go on to examine the characteristics of the firms utilizing continued patents, as well as the characteristics of the continued patents, broken down by the three types of continuation: CAP, CIP, and divisional.

Among their findings are the following:

- Patents in the "drugs and medicine" and the "chemicals" technology classes are among the most intensive users of continuations. For the years 1981–2000, 46 percent of patents issued were in drugs and medicines, and 36 percent were for chemicals.

- CIPs accounted for the majority of continuations in the drugs and medicine and chemicals technology classes.

- Hegde, Mowery, and Graham test whether the "pioneering" inventions of technology specialists (namely, small biotech companies) are more likely to issue from continuation by including an interaction factor in a model that captures the multiplicative effect of patent importance, R&D intensity, and patent intensity. In this case, a finding that continuation use is positively correlated with the combined effect of R&D intensity, patent intensity, and the quality of firm's patents supports the "pioneering inventor" characterization of continuation users.

- In the drug, medicine, and chemical industries, patents are counted as the most important mechanism for capturing value from innovation. The industries are also the most frequent users of CIPs. Patents utilizing CIPs contain significantly more backward citations, suggesting they are employed disproportionately to build upon inventors' prior work. In addition, CIPs produce patents that make 26 percent more claims than "ordinary" patents, reflecting their use by applicants to incorporate new material to support new claims. CIPs also receive a high number of forward citations within four years of their issue and are less likely to expire after four years than "ordinary" patents. "These results," the researchers conclude, "suggest that patents emerging from the CIPs are of higher private and technological 'value'" (Hegde, Mowery, and Graham 2007, 18).

- Overall, Hegde, Mowery, and Graham conclude, the continuation procedure for CIPs in the drug, medicine, and chemical industries supports "pioneering" inventors by allowing them to secure an early priority date and then revise their applications as the technology develops later. (They also point out that their research does not answer questions concerning submarine patenting strategies, which may also be associated with use of CIPS. As they note earlier in the paper, however, changes in patent term rules and publication of all patents after eighteen months are likely to curtail such activity substantially in the future.)

Though this research is not definitive, and the results are tentative in conjunction with the other arguments made above, it does, from our perspective, reinforce the case for caution and more analysis before plunging forward with major revisions to the continuation procedures.

The USPTO

No matter what the final disposition of the current legislation before Congress, or future changes proposed by administrative agencies or mandated by U.S. federal courts, the burden upon the USPTO and the entire patenting process in the United States is bound to increase dramatically— and this on top of responsibilities that, to many observers, are already overwhelming the system and the agency itself. It is, therefore, important for Congress and the executive to meet the challenges to the work of the USPTO with dispatch.

Among the changes we would endorse for the operations of the agency are the following:

- The fee system should be kept, but some mechanism should be put in place to assure that the agency retains all of the earnings from the application process and disperses them to improve the examination system, as mandated in pending legislation (H.R. 2336). In addition, Congress should reinstate a line item for the USPTO in the federal budget, after which it should determine what level of funding the agency needs to fulfill its old and new obligations and add the necessary funds to earnings from the fee system.

- At the same time, the patent office needs to examine its own internal procedures and modernize its operations. While this will include much greater utilization of information systems and computerized technology, it will also call for a reexamination of the way the office defines productivity and the incentives it creates for patent examiners to produce higher-quality patents. Despite the announcement with some fanfare of the 21st Century Plan to improve patent quality, top patent office management still places too much emphasis on "pleasing the customers" by measuring examiner productivity on the basis of how quickly examiners process patents and push them out the door rather than on their judgment of patent quality and their ability to reject bad patents and make the rejection stick (U.S. Patent and Trademark Office 2002). The compensation structure should be revamped to reward examiners for the quality of their "negative" judgments as well as for the number of patents they reward each year.

- That said, the total compensation package for examiners should be reviewed, with the aim of staunching the flow of examiners out of the patent office. As noted in chapter 6, a large percentage of young examiners leave after only a few years to assume higher-paying jobs in the private sector. A public institution can never match these opportunities fully, but means should be explored—through salary or other incentives—to retain experienced examiners.

- The experience of the last decade has underscored the vital necessity for the patent office to educate itself (and its examiners) on the latest trends and breakthroughs in a number of scientific and technological areas—particularly in biotechnology and electronics, but also in materials science and chemistry. A concerted effort should be made to attract and reward PhDs in strategic areas, and specialization among patent examiners should be encouraged. The USPTO should also establish more formal relations with scientific and engineering organizations and academic scientists and engineers. One possibility would be to enlist the aid of the National Academies of Science and Engineering in establishing a continuing formal relationship, possibly through an interdisciplinary advisory committee. (We would not, however, recommend giving that committee or advisory body any formal responsibilities over the patent office's examination procedures.)

8

Summary and Afterword

The pharmaceutical industry is exceptional in its reliance upon intellectual property. IP is essential for motivating both the initial discovery and post-discovery development of useful products. This is particularly true of the biotech pharmaceutical industry, where development times for initial approved uses are long, and where much of the most fruitful development occurs after initial FDA approval.

The scientific base undergirding biotech drug development continues to change with dizzying rapidity. Emerging results from the ENCODE project, for example, are altering views of the relationships between genes and protein expression, with potentially far-reaching implications for patents.[1] This is a useful reminder that what we know today about biotechnological drug development is but a snapshot of a process whose trajectory remains indistinct.

Patent law and the institutions surrounding it are also changing rapidly, partly because they must adjust to advances in science and in business methods (which have themselves evolved to take advantage of new technology). The relationships between specific industries and the dynamics of patent law are by no means consistent, however, as the effects of legal change on biotechnology are quite different from those on computer-based industries, or even the traditional small-molecule pharmaceutical industry.

All this has given rise to fundamental changes at the U.S. Patent and Trademark Office and an avalanche of litigation, with results that are altering important aspects of the patent system, including the treatment of IP in biotechnology and related industries. These developments and controversies have also produced legislative proposals to reform the patent system in major ways.

On the whole, the course of patent law in recent years reflects powerful self-correcting forces, as apparent excesses and miscalculations have been pulled back or reassessed, leaving most of the system operating reasonably well except for a lack of resources at the USPTO (including an inability to meet the challenges of new technological developments). The most feared developments have fortunately failed to occur, while both basic and applied research have proceeded with ever greater vigor. Although some legislative changes will be useful, there are compelling reasons for restraint. Given the rapidity and unpredictability of advances in basic science and their applications in biotechnology, along with parallel changes in patent law, Congress should approach patent reform with great caution. Large-scale changes may well have unintended negative consequences and could easily do more harm than good. On the other hand, limited changes aimed at making the granting and litigation of biotechnology patents and patents in other technology sectors more efficient could strengthen the patent system and foster greater innovation.

Notes

Chapter 1: Biotechnology and Health

1. Some useful sources on the history of the biotechnology pharmaceutical industry are studies by the Biotechnology Industry Organization (2005) and Ernst and Young (2003a; 2007), and the essays in *Biotechnology: Essays from Its Heartland* (Yarris 2004).

2. Gleevec was approved by the U.S. Federal Drug Administration in May 2001. Tommy Thompson, secretary of Health and Human Services, observed that "this single drug is as interesting and impressive as any we have ever seen throughout our long war on cancer" (Online NewsHour Update 2001). The pivotal clinical trial results are described by Druker et al. (2001). On treatment costs, see studies by Wade (2001) and Simon (2002).

3. The two articles that prompted this editorial are by Piccart-Gebhart (2005) and Romond et al. (2005).

4. We summarize this remarkable line of research in the section in chapter 3 on the costs and uncertainty of bringing new drugs to market; but see also Cross and Burmester (2006).

5. A series of publications from the Sabin Vaccine Institute describes both basic science and clinical work on therapeutic cancer vaccines; see Sabin Vaccine Institute (2004).

Chapter 2: The Sources of Biotechnology R&D

1. *Bayh-Doyle Act*, Public Law 96-517, 96th Cong., 1st sess. (December 12, 1980).

2. This history draws on Blumenthal's 2003 account.

3. A series of reports commissioned by the European Commission (coordinated by Italian economist Fabio Pammolli) offers many useful insights and data comparing European and American relationships between the private sector and academic or publicly supported biomedical research. See Allansdottir et al. (2002); Gambardella, Orsenigo, and Pammolli (2000); and Owen-Smith et al. (2002).

4. For 2005, the pharmaceutical industry trade group PhRMA reports $51.8 billion in R&D by member companies, some of which are primarily or

partly biotech firms (Pharmaceutical Research and Manufacturers of America 2007, 2). A substantial portion of the $20 billion in R&D that BIO reports for 2005 is not in the PhRMA total.

Chapter 3: Essential Features of the Biotechnology Industry

1. *Prescription Drug User Fee Act of 1992*, Public Law 102-571, 106 Stat. 4491, 102d Cong., 2d sess. (October 8, 1992).

2. On the Alzheimer's trial, see Tuszynski et al. (2005). On Parkinson's, see Stoessl (2007). Samakoglu et al. (2005) describe progress toward a gene therapy for sickle cell anemia and thalassemia, a closely related genetic disorder of red blood cells.

3. The *Wall Street Journal* (Hamilton 2002) described striking clinical trial results for Provenge, a therapeutic cancer vaccine from the biotech firm Dendreon. A series of publications from the Sabin Vaccine Institute describes both basic scientific and clinical work on therapeutic cancer vaccines; see Sabin Vaccine Institute (2004).

Chapter 4: The Role of Intellectual Property Rights

1. Mennel and Scotchmer (2005) provide an excellent recent review of the economics of intellectual property, including patents.

2. In chapter 5, we address several more refined arguments on the economic effects of patents. For more general surveys, see Hahn (2005) and Menell and Scotchmer (2005).

3. A useful summary of the theoretical and empirical literature is provided by Hahn (2005). See also Landes and Posner (2003).

4. Hahn (2005, 11) observes, "With a few notable exceptions (pharmaceuticals, for one), economists have been unable to show a clear causal link between increased patent rights and increased innovation."

5. We rely upon the summary in Hahn's 2005 study.

6. Again, we rely upon Hahn's (2005) summary.

7. For a description of various economic theories on the costs and benefits of patents and their role in the innovation process, see Mazzoleni and Nelson (1998) and Jaffe (2000).

8. Beyond the parameters of this study, the Ernst and Young annual survey of the biotechnology industry highlighted in 2007 the implications of the growing synergy between the biotechnology and pharmaceutical industries:

> There is no question that biotechnology is now the engine of innovation for the drug development industry. If there was any doubt about it before, the headline-grabbing mergers, acquisitions, and strategic

alliances of 2006 provide ample evidence of the tremendous potential latent in biotechnology's cutting-edge platforms, technologies and pipelines. In several large acquisitions, big pharma companies paid unprecedented premiums for early-stage biotech platforms . . . While much of this interest is fueled by big pharma's dwindling pipelines and large case reserves, it is also driven by another telling statistic—for several years in a row, biotech companies have secured more product approvals than their big pharma counterparts, even though big pharma significantly outspends the biotech industry on research and development. (Ernst and Young 2007, 1)

9. In a highly nuanced interpretation of the particular issues presented by biotechnology patents, two prominent legal scholars have argued for the courts to fashion policies—through case law—that incorporate the insights of Kitch's prospect theories while guarding against the allegedly negative consequences of the anticommons phenomenon. They maintain that the courts should hold biotech patents to a high standard of obviousness while lowering the standard for disclosure:

> This alternative approach—a fairly high obviousness threshold coupled with a fairly low disclosure requirement—will produce a few very powerful patents in uncertain industries. It will therefore solve the anticommons problem often identified with biotechnology, while at the same time boosting incentives to innovate. This calibration of patent frequency and scope seems to us the proper response to the anticommons concern found in much of the biotechnology literature. We worry that the alternate solution proposed by certain commentators, of largely eliminating biotechnology property rights in favor of governmental control over inventions supported by public funds, might unacceptably reduce the incentive for biotechnology companies to move beyond invention to innovation and product development. (Burk and Lemley 2003, 62)

In the past decade, however, the Court of Appeals for the Federal Circuit (CAFC; see chapter 5) has moved in the opposite direction, toward a lower standard of obviousness and stricter disclosure.

Chapter 5: Challenges to the Biotechnology Property Rights System

1. For an excellent introduction to many of the public policy issues surrounding intellectual property and human gene patents, including contributions by many interested parties and leading academics, see Korn and Heinig (2002).

2. In 2004, the FTC, the NAS, and the American Intellectual Property Law Association (AIPLA) jointly sponsored a series of town meetings on issues relating to the patent system and, in June 2005, they held a wrap-up conference in Washington, D.C. The 2004 NAS report was a project of the Committee on Intellectual Property Rights in a Knowledge-Based Economy, and the 2005 NAS report was a joint effort by the NRC Board on Science, Technology, and Economic Policy, and the NAS/NAE (National Academy of Engineering) Committee on Science, Technology, and Law.

3. *Madey v. Duke* 307 F.3d 1351 (Fed. Cir. 2002).

4. It should be noted, however, that Walsh, Cho, and Cohen's findings relating to impediments to the transfer of research material inputs told a somewhat different story. They found evidence that access to these inputs was somewhat restricted, and that some individual research projects were subject to delays or cancellation. On average, academics had made about seven requests for materials to other academics in the previous two years, and about two requests to industry labs. Some 19 percent did not receive materials in response to their latest requests. The researchers concluded that scientific competition and the cost and effort involved in complying with such requests were the main reasons for not fulfilling them—although for industry scientists, commercial interests were also important (Walsh, Cho, and Cohen 2005b, p. 2003).

5. For a survey and description of recent studies in this area, see McManis (2006).

6. The BRAC1 and BRAC2 genes are implicated in ovarian and breast cancer. The controversy developed over patent rights asserted by Myriad Genetic Laboratories for a diagnostic test utilizing BRAC1 and BRAC2 genes—and by the demand that others using the test pay a fee to Myriad. For more details, see Bunk (1999).

7. *Patent Act of 1952*, 66 Stat. 792, Pub. L. 82-593, 82d Congress, July 19, 1952.

8. *Brenner v. Manson*, 383 U.S. 519.

9. *Regents of the University of California v. Eli Lilly and Co.*, 119 F.3d. 1550 (Fed. Cir. 1997).

10. *University of Rochester v. G. D. Serle*, 375 F.3d 1303 (Fed. Cir. 2004).

11. *KSR International v. Teleflex Inc.*, No. 04-1350, 2007 (April 29, 2007).

12. *Diamond v. Chakrabarty*, 447 U.S. 303 (1980).

13. For a balanced analysis of the arguments for and against a "research exemption" for universities, see Thomas 2004.

14. Stephen Merrill, executive director of the NAS Science, Technology, and Economic Policy Board (STEP), has pointed out to us that both the 2004 and 2005 reports made legislative proposals; but we would still hold that the 2004 report was considerably more tentative in its analysis and recommendations (e-mail from Stephen Merrill to Claude Barfield, June 13, 2007).

15. Richard Nelson's proposals built upon a similar proposal by Rochelle Dreyfuss (2003). Dreyfuss would eliminate subsequent patent rights when a university received a research exemption for a project.

Chapter 6: The Drive for Legislative Solutions

1. *Omnibus Budget Reconciliation Act of 1990*, Public Law 101-508, 104 Stat. 1388, 101st Cong., 2d sess. (November 5, 1990).

2. It has been estimated that from 1994 through 2002, Congress withheld $573 million, although in recent years it has provided full funding for the office, and pending legislation—H.R. 2336—would mandate that in the future the USPTO retain all of its fee revenues (Kirk 2007).

3. The following analysis of the legislation will be based largely upon the legislation as originally introduced. Because of publication lags, we could not record changes that will occur in committee markups and floor action in the two houses of Congress.

4. Robert Merges, a preeminent legal scholar on intellectual property, described the challenges posed by the two industries to a "unitary" patent system at a joint conference of the NAS, the FTC, and the American Intellectual Property Law Association (AIPLA) in June 2005:

> One of the key questions I think that is in front of us . . . [is] whether or not the unitary system, which was created in the late 18th century, can still really work? We saw two very different perspectives. We had the software industry and the pharmaceutical industry, two really polar opposites in terms of how they use the patent system, and how patents work in those fields. In software, it's all about portfolios. Individual patents are not that important. And in many ways, their concerns about patents are: (a) generate licensing revenues; (b) use them defensively; and (c) try to fend off some of these extortionate trolls who are trying to come in and assert patents against them. Those issues find their way into the current reform bill that we're going to hear about later. Obviously, pharmaceuticals are different. Individual patents can be worth billions of dollars, the so-called billion-dollar-molecule patents. And the question for us today, which is really an extension of the old debate, is whether the patent system in its unitary form can handle these kinds of differential industries . . . I think that it's interesting to see in all the discussion of reform that nobody really has argued against going away from our uniform and unitary patent system. So I think the answer we have come to is yes, we can probably accommodate lots of different industries in lots of different ways. (National Academies Board on Science, Technology, and Economic Policy et al. 2005, 61–62)

5. PhRMA, the trade association of the U.S. pharmaceutical industry, has to date issued no formal statement on the legislation. But a number of large pharmaceutical companies are members of the 21st Century Coalition (see below), and general

pharma positions were confirmed by PhRMA staff in a telephone interview on June 1, 2007. In addition, it should be noted that a group of small biotech companies, state biotech associations, and several university technology-manager programs have banded together in an "innovation alliance" (Innovation Alliance 2007).

6. "Prior art" constitutes the existing body of technological information against which the patentability of an invention is evaluated.

7. See page 67 for explanation of "first-to-file."

8. Of course, the issue of patent reform has also sparked great interest among legal scholars. For a sharply argued and quite informative debate on many of the issues now before Congress, see the transcript of the panel discussion that took place at the Patent Scholars Conference, sponsored by the Santa Clara University High Technology Law Institute and the Berkeley Center for Law and Technology, in October 2006 (Barr et al. 2006). The panelists were Robert Barr, Berkeley Center of Law and Technology, University of California, Berkeley (moderator); Robert Merges, Boalt School of Law, University of California, Berkeley; Rebecca Eisenberg, University of Michigan Law School; and Kevin Outterson, West Virginia University College of Law.

Just as this study was going to press, U.S. labor unions raised the stakes regarding the pending legislation when they weighed in against many of the proposed changes, arguing they would undermine U.S. competitiveness in the struggle for world markets against rising developing countries such as China and India. In late July, also, a bipartisan group of 60 members of the House of Representatives, in a letter to the House party leaders, echoed these arguments and strongly recommended that Congress not rush to pass the patent reform legislation (Hitt 2007).

Chapter 7: Evaluation and Recommendations

1. This theme was expressed more colorfully by Rebecca Eisenberg, a prominent legal scholar highly critical of the current U.S. patent system, who stated, "We've seen Congress repeatedly try to implement changes that seem uncontroversial and do it in a half-assed way that leaves all these kind of dangling bits of statutory language that make the whole thing incoherent" (Barr et al. 2006, panel 5, 1).

2. In reaching our own conclusions on the legislation, we have utilized a variety of sources, but we have particularly benefited from the following studies and analyses: Jaffe and Lerner (2004); Dreyfuss (2006; review of Jaffe and Lerner); Maskus (2006); National Academy of Sciences (2004); and American Bar Association (2007).

3. A downside to retaining a grace period in a first-inventor-to-file system is that it reintroduces the possibility of fractious contests over who is the original inventor. Our American Enterprise Institute colleague Theodore Frank warned that the grace period would lead to a return, in some cases, to "messy factual

first-to-invent debates" (e-mail to the authors, June 15, 2007). A Congressional Research Service report also made this point, stating that retaining the grace period would mean that the U.S. shift from a first-to-invent to a first-inventor-to-file system would have been incomplete (Thomas and Schacht 2007b, CRS-18). Because the applicant's date of invention would have remained relevant, patentability decisions would have been more complex and less certain than in the first-inventor-to-file systems employed by all other patent-issuing states. The universities, however, have been adamant on this point; and, indeed, the U.S. government has pressed hard for this change in international patent harmonization negotiations.

4. *American Inventors Protection Act of 1999*, Public Law 106-113, 116 Stat. 1757-1922, 106th Cong., 1st sess. (November 26, 1999).

5. The rationale outlined here builds upon similar reasoning by Jaffe and Lerner (2004).

6. Joseph Farrell and Robert Merges, two legal scholars who have often criticized aspects of the U.S. patent system as tilting too far in favor of patent-holders, nonetheless have written in support of limiting the time period for administrative challenge:

> Post-grant patent revocations could be misused by firms who simply want to slow down or injure a patentee-firm . . . Safeguards must be built into the revocation system to prevent it from being overused. One response would be to limit patent revocations to some specific time period after the grant of a patent. This is far from ideal, given that the value of some patents will not be known (and hence the gains from invalidating these patents will not become clear) until well after patent issuance. Yet the general policy in the law of property favoring settled title argues for a cutoff to the post-grant challenge period. This will allow expectations regarding the value of the patent to settle, engendering commercial stability and fostering the market for patent licensing. (Farrell and Merges 2004, 967–68)

In the oft-cited National Academy of Sciences report (2004), the National Research Council committee that oversaw the process split on the issue of a "second window," with a majority opposed to it and a minority in favor.

7. *Lucent Technologies Inc. and Multimedia Patent Trust v. Gateway Inc. and Microsoft Corporation*, Case No. 02-CV-2060 B (CAB).

8. *Georgia-Pacific Corporation v. United States Plywood Corporation*, 318 F. Supp. 1116, 1120 (S.D.N.Y. 1970).

9. On May 3, 2007, the chief judge of CAFC, Paul Michel, took the highly unusual step of protesting to the chairman and ranking members of the Senate committee the proposed new rule on the apportionment of damages:

> The provision on apportioning damages would require courts to adjudicate the economic value of the entire prior art, the asserted patent claims and also all other features of the accused product or process whether or not patented. This is a massive undertaking for which courts are ill-equipped. For one thing, generalist judges lack experience and expertise in making such extensive, complex economic valuations, as do lay jurors. For another, courts would be inundated with massive amounts of data, requiring extra weeks of trial in nearly every case. Resolving the meaning of this novel language could take years . . . The provision also invites an unseemly battle of "hired-gun" experts opining on the basis of indigestible quantities of economic data . . . I am unaware of any convincing demonstration of the need for [this] provision. (Michel 2007, 1–2)

10. *eBay Inc. v. MercExchange, L.L.C.*, 126 S. Ct. 1837 (2006).

11. *Chevron, U.S.A. Inc. v. NRDC Inc.*, 467 U.S. 837 (U.S. 1984).

12. The American Intellectual Property Law Association also weighed in with strong opposition to the proposed changes in continuation rules. See Kirk (2006).

Chapter 8: Summary and Afterword

1. Specifically, the new research changes the traditional view that a collection of independent genes with each sequence of DNA links to a single function. Instead, it has now been demonstrated that genes appear to operate within a complex network and overlap and interact with each other and with other DNA elements in ways not yet understood. See Cookson (2007), Caruso (2007), and U.S. Department of Health and Human Services, National Human Genome Research Institute (2007).

References

Abboud, Leila. 2004. Lilly Expects Higher Profits Next Year. *Wall Street Journal*. December 9, B3.

Adams, Christopher P., and Van V. Brantner. 2006. Estimating the Costs of New Drug Development: Is It Really $802 Million? *Health Affairs* 25 (March 1): 420–28.

Adelman, David E., and Kathryn L. DeAngelis. 2007. Patent Metrics: The Mismeasure of Innovation in the Biotech Patent Debate. Arizona Legal Studies Discussion Paper No. 06-10. June 2007. http://ssrn.com/abstract=881842 (accessed June 20, 2007).

Agarwal, Rajshree, and Michael Gort. 2001. First-Mover Advantage and the Speed of Competitive Entry, 1887–1986. *Journal of Law and Economics* 44 (April): 161–77.

Allansdottir, Agnes, Andrea Bonaccorsi, Alfonso Gambardella, Myriam Mariani, Luigi Orsenigo, Fabio Pammolli, and Massimo Riccaboni. 2002. Innovation and Competitiveness in European Biotechnology. Enterprise Papers, No. 7, Enterprise Directorate-General, European Commission. http://ec.europa.eu/enterprise/library/enterprise-papers/pdf/enterprise_paper_07_2002.pdf (accessed July 23, 2007).

American Association for the Advancement of Science. 2005. Historical Data on Federal R&D, FY 1976–2006. Revised March 22. www.aaas.org/spp/rd/hist06p.pdf (accessed July 18, 2007).

American Bar Association. 2007. A Section White Paper: Agenda for 21st Century Patent Reform. May 1. http://www.abanet.org/intelprop/home/PatentReformWP.pdf (accessed June 25, 2007).

American Cancer Society. 2001. Angiogenesis Pioneer Is Focus of New Book and PBS Special: Early Funding Came from ACS. February 5. http://www.cancer.org/docroot/NWS/content/NWS_5_1x_Angiogenesis_Pioneer_Is_Focus_of_Book_and_PBS_Special.asp (accessed July 23, 2007).

American Intellectual Property Law Association. 2001. *Report of Economic Survey 2001*. Arlington, Va.: AIPLA.

Anand, Geeta. 2005. Company Puts a Big Price Tag on Drug to Treat Blood Disease. *Wall Street Journal*. December 29, D2.

Arora, Ashish, Andrea Fosfuri, and Alfonso Gambardella. 2001. *Markets for Technology: The Economics of Innovation and Corporate Strategy*. Cambridge, Mass.: MIT Press.

Association of American Universities, American Council on Education, National Association of State Universities and Land-Grant Colleges, Association of American Medical Colleges, and Council on Government Relations. 2007. *Comments on H.R. 1908 and S. 1145, the Patent Reform Act of 2007.* http://patentsmatter.com/media/issue/resources/20070501_UnivColl.pdf (accessed July 23, 2007).

Association of University Technology Managers (AUTM). 2002. *AUTM Licensing Survey: FY 2002*, ed. Ashley J. Stevens. http://www.autm.net/events/File/Surveys/02_Abridged_Survey.pdf (accessed June 25, 2007).

———. 2005a. Frequently Asked Questions. http://www.autm.net/aboutTT/aboutTT_faqs.cfm (accessed August 16, 2007).

———. 2005b. Now-Complete 2003 AUTM Licensing Survey Summary Details Canadian and U.S. Technology Transfer Trends. Press release. http://www.autm.net/news/dsp.newsDetails.cfm?nid=52 (accessed August 16, 2007).

———. 2006. Summary of U.S. Patent Reform Legislation Issues. October 6. http://www.autm.net/aboutTT/Patent_Reform_Legislation.pdf (accessed June 25, 2007).

Austin, David H. 2000. Patents, Spillovers and Competition in Biotechnology. Resources for the Future, Discussion Paper 0053. November. http://www.rff.org/rff/Documents/RFF-DP-00-53.pdf (accessed June 25, 2007).

Ball, Bret G., and W. French Anderson. 2000. Bringing Gene Therapy to the Clinic. *Journal of the American Medical Association* 284 (December 6): 2788–89.

Barinaga, Marcia. 2000. Angiogenesis Research: Cancer Drugs Found to Work in New Way. *Science* 288 (April 14): 245.

Barr, Robert, Robert Merges, Rebecca Eisenberg, and Kevin Outterson. 2006. Prospects for Patent Litigation in Congress. Panel transcript. Patent Scholars Conference: Patent Policy in the Supreme Court and Congress. Berkeley Center for Law and Technology and Santa Clara's Law and Technology Institute. October 27. http://tlf.vportal.net/generated/transcripts/transcript__807.html (accessed August 16,2007).

Begley, Sharon. 2005. Why Gene Therapy Still Hasn't Produced Forecast Breakthroughs. *Wall Street Journal.* February 18, B1.

Bessen, James E., and Robert Hunt. 2004. An Empirical Look at Software Patents. Federal Reserve Bank of Philadelphia Working Paper 03-17R. March 16. http://www.researchoninnovation.org/softpat.pdf (accessed July 24, 2007).

Biotechnology Industry Organization. 2005. BIO 2005–2006 *Guide to Biotechnology.* http://www.biotechwork.org/pages/FileStream.aspx?mode=Stream&fileId=84d27f43-4cf4-db11-b900-00c09f26cd10 (accessed August 1, 2007).

———. 2006. The Comments of the Biotechnology Industry Organization on the United States Patent and Trademark Office: Proposed Rules Changes Concerning Continuation Practice and Claim Limitations. *Biotechnology Law Report* 25 (4): 473–83. May 2. http://www.liebertonline.com/doi/abs/10.1089/blr.2006.25.473?cookieSet=1&journalCode=blr (accessed August 1, 2007).

————. 2007a. BIO Expresses Concerns Regarding New Patent Reform Legislation. April 18. http://www.bio.org/news/newsitem.asp?id=2007_0418_06 (accessed June 25, 2007).

————. 2007b. *Guide to Biotechnology 2007*. http://www.bio.org/speeches/pubs/er/BiotechGuide.pdf (accessed August 14, 2007).

————. n.d. Biotechnology Industry Facts. http://bio.org/speeches/pubs/er/statistics.asp (accessed July 18, 2007).

Blumenthal, David. 2003. Academic-Industrial Relationships in the Life Sciences. *New England Journal of Medicine* 349 (25): 2452–59, December 18.

Brendza, Robert P., Brian J. Bacskai, John R. Cirrito, Kelly A. Simmons, Jesse M. Skoch, William E. Klunk, Chester A. Mathis, Kelly R. Bales, Steven M. Paul, Bradley T. Hyman, and David M. Holtzman. 2005. Anti-A, Antibody Treatment Promotes the Rapid Recovery of Amyloid-Associated Neuritic Dystrophy in PDAPP Transgenic Mice. *Journal of Clinical Investigation* 115 (2): 428–33.

Brody, Baruch A. 1995. *Ethical Issues in Drug Testing, Approval, and Pricing*. New York: Oxford University Press.

Bunk, Steve. 1999. Researchers Feel Threatened by Disease Gene Patents. *Scientist*. October, 7.

Burk, Dan L., and Mark A. Lemley. 2003. Biotechnology's Uncertainty Principle. *University of Minnesota School of Law Public Law and Technology Review*. Research Paper No. 03-5: 62. http://papers.ssrn.com/sol3/papers.cfm?abstract_id=303619 (accessed August 8, 2007).

Bush, Vannevar. 1945. *Science the Endless Frontier: A Report to the President by Vannevar Bush, Director of the Office of Scientific Research and Development, July 1945*. http://www.nsf.gov/od/lpa/nsf50/vbush1945.htm (accessed January 2, 2006).

Calfee, John. 2007a. The Golden Age of Medical Innovation. *The American: A Magazine of Ideas*. March/April, 41–52.

————. 2007b. Lessons of the Heart. *The American: A Magazine of Ideas* April 3. http://american.com/archive/2007/april-0407/lessons-of-the-heart/?searchterm=Lessons%20of%20the%20Heart (accessed August 1, 2007).

Calfee, John E., and Elizabeth DuPré. 2006. The Emerging Market Dynamics of Targeted Therapeutics. *Health Affairs* 25 (5): 1302–8.

Caruso, Denise. 2007. A Challenge to Gene Theory, a Tougher Look at Biotech. *New York Times*. July 1.

Caulfield, Timothy, Robert M. Cook-Deegan, F. Scott Kieff, and John P. Walsh. 2006. Evidence and Anecdotes: An Analysis of Human Gene Patenting Controversies. *Nature Biotechnology* 24 (September): 1091–94.

Choudhry, Niteesh K., Jerry Avorn, Elliott M. Antman, Sebastian Schneeweiss, and William H. Shrank. 2007. Should Patients Receive Secondary Prevention Medications for Free after a Myocardial Infarction? An Economic Analysis. *Health Affairs* 26 (1): 186–94.

Coalition for 21st Century Patent Reform. 2007. Views on the "Patent Reform Act of 2007" (S. 1145/H.R. 1908). April 27. http://www.ipfrontline.com/depts/article.asp?id=14890&deptid=4&page=1 (accessed August 1, 2007).

Cohen, Wesley J., and Richard C. Levin. 1989. Empirical Studies of Innovation and Market Structure. In *Handbook of Industrial Organization*, ed. Richard Schmalensee and Robert Willig. Amsterdam: North Holland.

Cohen, Wesley J., Richard R. Nelson, and John P. Walsh. 2000. Protecting Their Intellectual Assets: Appropriability Conditions and Why U.S. Manufacturing Firms Patent (or Not). Working paper. Carnegie Mellon University. January.

Cookson, Clive. 2007. Research Reveals Complexity in How Human Genes Interact. *Financial Times* (London). June 14.

Cross, Deanna, and James K. Burmester. 2006. Gene Therapy for Cancer Treatment: Past, Present, and Future. *Clinical Medicine and Research* 4 (3): 218–27.

Cutler, David. 2001. Declining Disability among the Elderly. *Health Affairs* 20 (6): 11–27.

Daiger, Stephen P. 2005. Was the Human Genome Project Worth the Effort? *Science* 308 (April 15): 362–63.

Danzon, Patricia. 1998. Economics of Parallel Trade. *Pharmacoeconomics*. March.

Danzon, Patricia M., and Adrian Towse. 2002. The Economics of Gene Therapy and of Pharmacogenomics. *Value in Health* 5 (1): 5–13.

Datamonitor. 2002. Xigris: Lilly's Sepsis Flop Misses Out. March. www.inpharm.com/intelligence/datamonitor040302.html (accessed July 5, 2002).

Desrosiers, Ronald L. 1989. The Drug Patent Term: Longtime Battleground in the Control of Health Care Costs. *New England Law Review* 24 (Fall): 133–34.

Dickson, Michael, and Jean Paul Gagnon. 2004. Key Factors in the Rising Cost of New Drug Discovery and Development. *Nature Reviews Drug Discovery* 3 (5): 417–29.

DiMasi, Joseph A., Ronald W. Hansen, and Henry G. Grabowski. 2003. The Price of Innovation: New Estimates of Drug Development Costs. *Journal of Health Economics* 22:151–85.

DiMasi, Joseph A., and Cherie Paquette. 2004. The Economics of Follow-On Drug Research and Development: Trends in Entry Rates and Timing of Development. *Pharmacoeconomics* 22 (supp. 2): 1–14.

Dooren, Jennifer Corbett. 2005. Femara Receives Wider Approval for Breast Cancer. *Wall Street Journal*. December 29.

Dreyfuss, Rochelle. 2003. Varying the Course. In *Perspectives on Properties of the Human Genome Project*, ed. F. Kieff. New York: Academic Press.

———. 2006. Pathological Patenting: The PTD as Cause or Cure. *Michigan Law Review* 104 (April): 1559–78.

Druker, Brian J., Charles L. Sawyers, Hagop Kantarjian, Debra J. Resta, Sofia Fernandes Reese, John M. Ford, Renaud Capdeville, and Moshe Talpaz. 2001. Activity of a Specific Inhibitor of the BCR-ABL Tyrosine Kinase in the Blast Crisis of Chronic Myeloid Leukemia and Acute Lymphoblastic Leukemia with the Philadelphia Chromosome. *New England Journal of Medicine* 344 (14): 1038–42.

Economist. 2007. Patently Obvious. May 5, 78.

Eisenberg, Rebecca S. 1997. Patenting Research Tools and the Law. In *Intellectual Property Rights and Research Tools in Molecular Biology: Summary of a Workshop Held at the National Academy of Sciences, February 15–16, 1996.* Washington, D.C.: National Research Council, National Academy Press. http://www. nap.edu/reading room/books/property/ (accessed August 1, 2007).

————. 2002. Why the Gene Patenting Controversy Persists. *Academic Medicine* 77 (12): 1381–87.

Ernst and Young. 2003. *Resilience: Americas Biotechnology Report 2003.* July. http://biozine.kribb.re.kr/kboard_trend/download.php?code=policy&no=1601 (accessed August 14, 2007).

————. 2007. *Beyond Borders: Ernst and Young's Global Biotechnology Report 2007.* Boston: Ernst and Young Global Biotechnology Center.

Farrell, Joseph, and Robert P. Merges. 2004. Incentives to Challenge and Defend Patents: Why Litigation Won't Reliably Fix Patent Office Errors and Why Administrative Patent Review Might Help. *Berkeley Technology Law Journal* 19 (Summer): 943–69.

Fedoroff, Nina, and Nancy Marie Brown. 2004. *Mendel in the Kitchen: A Scientist's View of Genetically Modified Foods.* Washington, D.C.: Joseph Henry Press.

Ferrara, Napoleone. 2002. VEGF and the Quest for Tumour Angiogenesis Factors. *Nature Reviews Cancer* 2:795–803.

Fisher, Lawrence M. 1999. A Vitalizing Gene, Straight to the Heart. *New York Times.* August 29.

Fleischer-Black, Matt. 2003. Wake-Up Call. *IP Law and Business.* October 6.

Ford, Earl S., Umed A. Ajani, Janet B. Croft, Julia A. Critchley, Darwin R. Labarthe, Thomas E. Kottke, Wayne H. Giles, and Simon Capewell. 2007. Explaining the Decrease in U.S. Deaths from Coronary Disease, 1980–2000. *New England Journal of Medicine* 356 (June 7): 2388–98.

Fox, N. C., R. S. Black, S. Gilman, M. N. Rossor, S. G. Griffith, L. Jenkins, and M. Koller. 2005. Effects of A Immunization (AN1792) on MRI Measures of Cerebral Volume in Alzheimer Disease. *Neurology* 64:1563–72.

Freedman, Vicki A., Linda G. Martin, and Robert F. Schoeni. 2002. Recent Trends in Disability and Functioning Among Older Adults in the United States: A Systematic Review. *Journal of the American Medical Association* 288 (24): 3137–46.

Fries, James F. 2002. Reducing Disability in Older Age. *Journal of the American Medical Association* 288 (24): 3164–66.

Gambardella, Alfonso, Luigi Orsenigo, and Fabio Pammolli. 2000. *Global Competitiveness in Pharmaceuticals: A European Perspective.* Report prepared for the Enterprise Directorate-General of the European Commission. November. http://ec.euro pa.eu/enterprise/library/enterprise-papers/pdf/enterprise_paper_01_2001.pdf (accessed August 15, 2007).

Gilbert, Jim, Preston Henske, and Ashish Singh. 2003. Rebuilding Big Pharma's Business Model. *In Vivo: The Business & Medicine Report* 21 (10): 73.

Golden, John. 2001. Biotechnology, Technology Policy, and Patentability. *Emory Law Journal* 101:34–35.

Grabowski, Henry. 2002. Patents and New Product Development in the Pharmaceutical and Biotechnology Industries. Paper presented at the Science and Cents: Exploring the Economics of Biotechnology Conference, Federal Reserve Bank of Dallas. April. http://www.dallasfed.org/research/pubs/science/index.html (accessed August 8, 2007).

———. 2007. Data Exclusivity for New Biological Entities. Working paper. Duke University Department of Economics. June.

Grabowski, Henry G., and John Vernon. 2000. New Findings on the Returns to Pharmaceutical Research and Development. *Pharmacoeconomics* 18 (supp. 1): 21–32.

Green, J., and S. Scotchmer. 1995. On the Division of Profit in Sequential Innovation. *RAND Journal of Economics* 26 (Spring): 20–33.

Greenhouse, Linda. 2007. High Court Puts Limits On Patents. *New York Times.* May 1, C1, C6.

Greil, Anita (Dow Jones Newswires). 2005. EU Approves the Wider Use Of Roche Cancer Drug Xeloda. *Wall Street Journal.* April 4.

Hahn, Robert W. 2005. An Overview of the Economics of Intellectual Property Protection. In *Intellectual Property Rights in Frontier Industries: Software and Biotech*, ed. Robert W. Hahn. Washington, D.C.: AEI-Brookings Joint Center for Regulatory Studies.

Hall, Bronwyn H., Stuart J. H. Graham, Dietmar Harhoff, and David C. Mowery. 2003. Prospects for Improving U.S. Patent Quality via Post-Grant Opposition. NBER Working Paper 9731. National Bureau of Economic Research. May.

Hamilton, David P. 2002. Dendreon Prostate-Cancer Drug Helps Only a Subgroup of Patients. *Wall Street Journal.* August 9.

Hardin, Garrett. 1968. The Tragedy of the Commons. *Science* 162:1243.

Hedlund, Julie A. 2007. Patents Pending: Patent Reform for the Innovation Economy. Information Technology & Innovation Foundation. May. http://www.itif.org/files/PatentsPending.pdf (accessed June 22, 2007).

Hegde, Deepak, David Mowery, and Stuart Graham. 2007. Pioneers, Submarines, or Thickets: Which Firms Use Continuations in Patenting and Why? Draft. February 12. http://elsa.berkeley.edu/~bhhall/e222spring07_files/HegdeMoweryGraham_11Feb07.pdf (accessed June 25, 2007).

Heller, Michael. 1998. The Tragedy of the Anticommons: Property in the Transition from Marx to Markets. *Harvard Law Review* 111:621–88.

Heller, Michael, and Rebecca Eisenberg. 1998. Can Patents Deter Innovation? The Anticommons in Biomedical Research. *Science* 280 (5364): 698–701. http://www.sciencemag.org/cgi/content/full/280/5364/698 (accessed June 19, 2007).

High, Katherine. 2005. Anemia and Gene Therapy: A Matter of Control. *New England Journal of Medicine* 352 (11): 1146–47.

Hitt, Greg. 2007. Patent System's Revamp Hits the Wall. *Wall Street Journal.* August 27.

Holman, Christopher M. 2006. Biotechnology's Prescription for Patent Reform. *John Marshall Review of Intellectual Property Law* 5 (Spring): 318–47.

Hortobagyi, Gabriel N. 2005. Trastuzumab in the Treatment of Breast Cancer. *New England Journal of Medicine* 353 (October 20): 1734–36.

InfoService Biotechnology. 2007. Sales of Recombinant Drugs, 2005. http://www.i-s-b.org/business/rec_sales.htm (accessed August 3, 2007).

Innovation Alliance. 2007. Not Patently Obvious: The Innovation Alliance's Position on Proposed Patent Reform Legislation. March. http://www.innovation alliance.net/resource_center/pdfs/IA_Patent_Reform_White_Paper.pdf (accessed June 25, 2007).

Jaffe, Adam B. 2000. The U.S. Patent System in Transition: Policy Innovation and the Innovation Process. *Research Policy* 29: 531–57.

Jaffe, Adam B., and Josh Lerner. 2004. *Innovation and Its Discontents: How Our Broken Patent System Is Endangering Innovation and Progress, and What to Do About It.* Princeton, N.J.: Princeton University Press.

———. 2006. Innovation and Its Discontents. *Capitalism and Society* 1 (3): article 3. http://www.bepress.com/cas/vol1/iss3/art3/ (accessed June 22, 2007).

Jain, Rakesh K. 2005. Normalization of Tumor Vasculature: An Emerging Concept in Antiangiogenic Therapy. *Science* 307 (January 7): 58–62.

Johnson, Matthew, Wesley Cohen, and Brian Junker. 1999. Measuring Appropriability Research and Development with Item Response Models. Technical Report No. 690. Department of Statistics, Carnegie Mellon University. February. http://www.stat.cmu.edu/tr/tr690/tr690.pdf (accessed July 25, 2007).

Kaiser, Jocelyn. 2005. Putting the Fingers on Gene Repair. *Science* 310 (December 23): 1890–91.

Kirk, Michael K. 2006. Letter to the Honorable Jon Dudas (re: 71 Fed. Reg. 48, January 3, 2006). April 24. http://www.aipla.org/Template.cfm?template=/ContentManagement/ContentDisplay.cfm&ContentID=11276 (accessed August 1, 2007).

———. 2007. Letter to Representatives John Conyers, Howard Berman, Lamar Smith, and Howard Coble. May 28. http://www.aipla.org/Content/Content Groups/Legislative_Action/110th_Congress1/Testimony6/LtrReHR2336.pdf (accessed August 8, 2007).

Kitch, Edmund. 1977. The Nature and Function of the Patent System. *Journal of Law and Economics* 20 (October): 265–90.

Kleinke, J. D. 2001. The Price of Progress: Prescription Drugs in the Health Care Market. *Health Affairs* 25 (5): 43–60.

Korn, David, and Stephen J. Heinig, eds. 2002. Special Theme Issue: Public Versus Private Ownership of Scientific Discovery: Legal and Economic Analyses of the Implications of Human Gene Patents. Special issue, *Academic Medicine* 77, no. 12.

Kortum, Samuel, and Josh Lerner. 1999. Stronger Protection or Technological Revolution: What Is Behind the Recent Surge in Patenting? NBER Working Paper 6204. http://www.nber.org/papers/w6204 (accessed August 16, 2007).

————. 2003. Unraveling the Patent Paradox. Unpublished working paper. University of Minnesota Department of Economics and Harvard Business School.

Krasner, Jeffrey. 2004. Change of Habit Might Be Key to Drug's Success: MS Patients Must Visit Doctor for a Monthly Infusion. *Boston Globe*. November 15.

Lakdawalla, Darius, and Tomas Philipson. 1999. Aging and the Growth of Long-Term Care. Harris School Working Paper 99.5. University of Chicago.

Landes, William M., and Richard A. Posner. 2003. *The Economic Structure of Intellectual Property Law.* Cambridge, Mass.: The Belknap Press of Harvard University Press.

Lawrence, Stacy. 2007. Tech Transfer Fails to Translate into Startups. *Nature Biotechnology* 25: 616.

Lednicer, Dan. 2002. Tracing the Origins of COX-2 Inhibitors' Structures. *Current Medicinal Chemistry* 9 (15): 1457–61.

Lemley, Mark. 2001. Rational Ignorance at the Patent Office. *Northwestern University Law Review* 95 (4): 1495–1532.

Lemley, Mark, and Colleen Chien. 2003. Are the U.S. Patent Priority Rules Necessary? Boalt Working Papers in Public Law, No. 32. http://repositories.cdlib.org/boaltwp/32 (accessed June 22, 2007).

Lemley, Mark, and Kimberly Moore. 2004. Ending Abuse of Patent Continuations. *Boston University Law Review* 84 (February): 63–124.

Lerner, Joshua. 1994. The Importance of Patent Scope: An Empirical Analysis. *RAND Journal of Economics* 25 (2): 319–33.

Levin, Richard C., Alvin K. Klevorick, Richard R. Nelson, and Sidney G. Winter. 1987. Approaching the Returns from Industrial R&D. *Brookings Papers on Economic Activity* 3:783–831.

Lichtenberg, Frank R. 1996. Do (More and Better) Drugs Keep People Out of Hospitals? *American Economic Review* 86 (May): 384–88.

————. 2002a. Benefits and Costs of Newer Drugs: An Update. NBER Working Paper 8996. National Bureau of Economic Research. June.

————. 2002b. The Economic Benefits of New Drugs. *Economic Realities in Health Care Policy* 2 (2): 18.

————. 2003. The Impact of New Drug Launches on Longevity: Evidence from Longitudinal Disease-Level Data from 52 Countries, 1982–2001. NBER Working Paper 9754. National Bureau of Economic Research. June.

Lichtenberg, Frank R., and Tomas J. Philipson. 2002. The Dual Effects of Intellectual Property Regulations: Within- and Between-Patent Competition in the U.S. Pharmaceuticals Industry. *Journal of Law and Economics* 45 (October): 643–72.

Maebius, Stephen B., and Harold C. Wegner. 2002. Ruling on Research Exemption Roils Universities. *National Law Journal.* December 16.

Malakoff, David. 2004. NIH Roils Academe with Advice on Licensing DNA Patents. *Science* 303 (March 19): 1757–58.

Mansfield, Edwin. 1986. Patents and Innovation: An Empirical Study. *Management Science* 32 (2): 173–81.

Marcus, Amy Dockser. 2004. Cancer Survivors Grapple With Issues Beyond Their Health. *Wall Street Journal.* November 30.

Marshall, Eliot. 2002. Cancer Therapy: Setbacks for Endostatin. *Science* 295 (March 22): 2198–99.

Marx, Jean. 2005. Encouraging Results for Second-Generation Antiangiogenesis Drugs. *Science* 308 (May 27): 1248–49.

Maskus, Keith E. 2006. Reforming U.S. Patent Policy: Getting the Incentives Right. Council on Foreign Relations, Council Special Report (CSR) No. 19. November. http://www.cfr.org/content/publications/attachments/PatentCSR.pdf (accessed July 25, 2007).

Matthay, Michael A. 2001. Severe Sepsis—A New Treatment with Both Anticoagulant and Antiinflammatory Properties. *New England Journal of Medicine* 334 (10): 759–61.

Mazzoleni, Roberto, and Richard R. Nelson. 1998. Economic Theories about the Benefits and Costs of Patents. *Journal of Economic Issues* 32 (4): 1031–52

McManis, Charles. 2006. The Impact of the Bayh-Dole Act on Genetic Research and Development: Evaluating the Arguments and Empirical Evidence to Date. Paper presented at the Patent Scholars Conference, jointly sponsored by the Santa Clara University High Technology Law Institute and the Berkeley Center for Law and Technology. October 26.

Menell, Peter, and Suzanne Scotchmer. 2007. Intellectual Property Law. In *Handbook of Law and Economics*, vol. 2, ed. A. Mitchell Polinsky and Steven Shavell. Amsterdam: North Holland Press, forthcoming. Prepublication text available at http://papers.ssrn.com/sol3/papers.cfm?abstract_id=741424 (accessed August 16, 2007).

Merges, Robert P., and Richard R. Nelson. 1990. On the Complex Economics of Patent Scope. *Columbia Law Review* 90 (May): 839–916.

Michel, Paul R. (chief judge, U.S. Court of Appeals for the Federal Circuit). 2007. Letter to Senators Patrick Leahy and Orrin Hatch. May 3. http://www.patentbaristas.com/wp/wp-content/uploads/2007/05/michellettermay3rd.pdf (accessed August 1, 2007).

Minna, John D., Adi F. Gazdar, Stephen R. Sprang, and Joachim Herz. 2004. A Bull's Eye for Targeted Lung Cancer Therapy. *Science* 304 (5676): 1458–61.

Mitchell, Peter. 2004. Erbitux Diagnostic Latest Adjunct to Cancer Therapy. *Nature Biotechnology* 22:363–64.

Mossinghoff, Gerald J. 2002. The First-to-Invent System Has Provided No Advantage to Small Entities. *Journal of the Patent and Trademark Office Society* 84 (6): 425–30.

Munro, Neil, and Andrew Noyes. 2007. Patent Reform, Pending. *National Journal.* May 5, 56–58.

Murphy, Kevin, and Robert Topel. 1999. *The Economic Value of Medical Research*. Chicago: University of Chicago Press.

Murray, Fiona. 2007. The Stem-Cell Market—Patents and the Pursuit of Scientific Progress. *New England Journal of Medicine* 356 (23): 2341–43.

National Academies Board on Science, Technology and Economic Policy, American Intellectual Property Law Association, and U.S. Federal Trade Commission. 2005. Transcript. Conference on Patent Reform. Washington, D.C. June 9. http://www.aipla.org/Content/ContentGroups/Meetings_and_Events1/Roadshows/20058/Transcript_6-9-05.pdf (accessed June 25, 2007).

National Academy of Sciences. 2003. *Patents in the Knowledge-Based Economy*, ed. Wesley M. Cohen and Stephen A. Merrill. Washington, D.C.: National Academies Press. Full text available at http://www.nap.edu/catalog/10770.html#toc (accessed July 27, 2007).

———. 2004. *A Patent System for the 21st Century*, ed. Stephen Merrill, Richard C. Levin, and Mark B. Myers. Washington, D.C.: National Academies Press. Full text available at http://www.nap.edu/catalog/10976.html#toc (accessed July 27, 2007).

———. 2005. *Reaping the Benefits of Genomic and Proteomic Research: Intellectual Property Rights, Innovation and Public Health*, ed. Stephen A. Merrill and Anne-Marie Mazza. Washington, D.C.: National Academies Press. Full text available at http://www.nap.edu/catalog/11487.html (accessed July 27, 2007).

National Science Foundation. 2005. Increase in U.S. Industrial R&D Expenditures Reported for 2003 Makes Up for Earlier Decline. By Raymond M. Wolfe. *SRS Info-Brief*. NSF 06-305. December. http://www.nsf.gov/statistics/infbrief/nsf06305/nsf06305.pdf (accessed July 26, 2007).

———. 2007. U.S. R&D Increased 6.0% in 2006 according to NSF Projections. By Brandon Shackelford. *SRS InfoBrief*. NSF 07-317. April. http://www.nsf.gov/statistics/infbrief/nsf07317/nsf07317.pdf (accessed July 26, 2007).

Nature Biotechnology. 2007. Burning Bridges. 25 (1): 2.

Nelson, Richard R. 2003. The Market Economy and the Scientific Commons. LEM Papers Series, No. 2003/24. Laboratory of Economics and Management, Sant'Anna School of Advanced Studies, Pisa, Italy. http://www.lem.sssup.it/WPLem/files/2003-24.pdf (accessed August 8, 2007).

Neyt, M., J. Albrecht, and V. Cocquyt. 2006. An Economic Evaluation of Herceptin in Adjuvant Setting: The Breast Cancer International Research Group 006 Trial. *Annals of Oncology* 17 (March): 381–90.

Nicoll, James A. R., David Wilkinson, Clive Holmes, Phil Steart, Hannah Markham, and Roy O. Weller. 2003. Neuropathology of Human Alzheimer Disease after Immunization with Amyloid-μ Peptide: A Case Report. *Nature Medicine* 9 (March): 448–552.

Online NewsHour Update. 2001. FDA Approves New Cancer Drug. May 10. http://www.pbs.org/newshour/bb/health/jan-june01/cancerupdate_5-10.html (accessed August 1, 2007).

Owen-Smith, Jason, Massimo Riccaboni, Fabio Pammolli, and Walter W Powell. 2002. A Comparison of U.S. and European University-Industry Relations in the Life Sciences. *Management Science* 48 (1): 24–43.

Pearson, Helen. 2006. When Good Cholesterol Turns to Bad. *Nature* 444 (7121): 794–95.

Pharmaceutical Research and Manufacturers of America (PhRMA). 2002. *Pharmaceutical Industry Profile 2002.* Washington, D.C.: PhRMA.

———. 2003. *Pharmaceutical Industry Profile 2003.* Washington, D.C.: PhRMA.

———. 2007. *Industry Profile 2007.* Washington, D.C.: PhRMA. http://www. phrma .org/files/Profile%202007.pdf (accessed August 1, 2007).

Philipson, Lennart. 2005. Medical Research Activities, Funding, and Creativity in Europe. *Journal of the American Medical Association* 294 (11): 1394–98.

Piccart-Gebhart, Martine J. 2005. Trastuzumab after Adjuvant Chemotherapy in HER2-Positive Breast Cancer. *New England Journal of Medicine* 353 (October 20): 1659–72.

Pimentel, David. 2004. Changing Genes to Feed the World. A review by David Pimentel of *Mendel in the Kitchen: A Scientist's View of Genetically Modified Foods*, by Nina Fedoroff and Nancy Marie Brown. *Science* 306 (October 29): 815.

Pollack, Andrew. 2005. New Genome Project to Focus on Genetic Links with Cancer. *New York Times.* December 14.

———. 2007. 3 Patents on Stem Cells Are Revoked in Initial Review. *New York Times.* April 3, C2.

Rai, Arti K. 2001. Fostering Cumulative Innovation in the Biopharmaceutical Industry: The Role of Patents and Antitrust. *Berkeley Technology Law Journal* 16:813–53.

Rai, Arti K., and Rebecca S. Eisenberg. 2003. Bayh-Dole Reform and the Progress of Biomedicine: Conference on the Public Domain. *Law and Contemporary Problems* 66 (Winter-Spring): 289–314.

Regalado, Antonio. 2003. To Sell Pricey Drug, Eli Lilly Fuels a Debate Over Rationing. *Wall Street Journal.* September 18.

Reichert, Janice M., and Christopher Milne. 2002. Public and Private Sector Contributions to the Discovery and Development of "Impact" Drugs. White paper. Tufts Center for the Study of Drug Development. May.

Romond, Edward H., Edith A. Perez, John Bryant, Vera J. Suman, Charles E. Geyer, Nancy E. Davidson, et al. 2005. Trastuzumab Plus Adjuvant Chemotherapy for Operable HER2-Positive Breast Cancer. *New England Journal of Medicine* 353 (October 20): 1673–84.

Rosen, Fred S. 2002. Editorial: Successful Gene Therapy for Severe Combined Immunodeficiency. *New England Journal of Medicine* 346 (16): 1241–43.

Ross, Susan D., I. Elaine Allen, Janet E. Connelly, Bonnie M. Korenblat, M. Eugene Smith, Daren Bishop, and Don Luo. 1999. Clinical Outcomes in Statin Treatment Trials. *Archives of Internal Medicine* 159 (15): 1793–1802.

Sabin Vaccine Institute. 2004. Proceedings of the Sixth Annual Walker's Cay Colloquium on Cancer Vaccines and Immunotherapy. http://www.sabin.org/files/PDF/wc2001.pdf (accessed July 10, 2007).

Samakoglu, Selda, Leszek Lisowski, Tulin Budak-Alpdogan, Yelena Usachenko, Santina Acuto, Rosalba Di Marzo, Aurelio Maggio, et al. 2005. A Genetic Strategy to Treat Sickle Cell Anemia by Coregulating Globin Transgene Expression and RNA Interference. *Nature Biotechnology* 24:89–94.

Saul, Stephanie. 2005. Bristol-Myers Squibb Receives U.S. Approval for Arthritis Drug. *New York Times*, December 24.

Schacht, Wendy H. 2000. *Federal R&D, Drug Discovery, and Pricing: Insights from the NIH-University-Industry Relationship.* CRS Report for Congress. Congressional Research Service. June 19. http://www.law.umaryland.edu/marshall/crsreports/crsdocuments/RL30585.pdf (accessed July 26, 2007).

———. 2006. *Patent Reform: Issues in the Biomedical and Software Industries.* CRS Report for Congress. Congressional Research Service. April 7. http://ftp.fas.org/sgp/crs/misc/RL33367.pdf (accessed July 26, 2007).

Schacht, Wendy H., and John R. Thomas. 2005. *Patent Reform: Innovation Issues.* CRS Report for Congress. Congressional Research Service. July 15. http://www.lawand innovation.org/cli/documents/crs_report_patent_reform.pdf (accessed July 26, 2007).

———. 2006. *Patent Reform: Innovation Issues.* CRS Report for Congress. Congressional Research Service. October 16 (update). http://ipmall.info/hosted_resources/crs/RL32996-061018.pdf (accessed July 26, 2007).

Schenk, Dale, Robin Barbour, Whitney Dunn, Grace Gordon, Henry Grajeda, Teresa Guido, Kang Hu, et al. 1999. Letters to Nature: Immunization with Amyloid-μ Attenuates Alzheimer-Disease-Like Pathology in the PDAPP Mouse. *Nature* 400 (July): 173–77.

Scherer, Frederic M. 2002. The Economics of Human Gene Patents. *Academic Medicine* 77 (12): 1348–67.

———. 2006. The Political Economy of Patent Policy Reform in the United States. Working Paper 06-22. AEI-Brookings Joint Center for Regulatory Studies. October. http://www.aei-brookings.org/publications/abstract.php?pid=1123 (accessed June 22, 2007).

Scotchmer, S. 1991. Standing on the Shoulders of Giants: Cumulative Research and Patent Law. *Journal of Economic Perspectives* 5 (Winter): 29–41.

Seiden, Carl. 2001. Zovant (for Sepsis)—A Potential Blockbuster. Report. New York: J. P. Morgan Securities Inc.

Sela, Michael, and Maurice R. Hilleman. 2004. Therapeutic Vaccines: Realities of Today and Hopes for Tomorrow. *Proceedings of the National Academy of Sciences* 101 (supp. 2): 14559.

Simon, Francois. 2002. Case Study, Gleevec: Success by Design in Oncology. Columbia University Business School.

Sipress, Alan. 2007. Patently at Odds. *Washington Post*. April 18, D1, D3.

Soumerai S. B., T. J. McLaughlin, D. Ross-Degnan, C. S. Casteris, P. Bollini. 1994. Effects of Limiting Medicaid Drug-Reimbursement Benefits on the Use of Psychotropic Agents and Acute Mental Health Services by Patients with Schizophrenia. *New England Journal of Medicine* 331:650–55.

Spilker, Bert. 1994. *Multinational Pharmaceutical Companies: Principles and Practices*. 2d ed. New York: Raven Press.

Steinberg, Daniel. 2006. An Interpretive History of the Cholesterol Controversy, Part V: The Discovery of the Statins and the End of the Controversy. *Journal of Lipid Research* 47 (July): 1339–51.

Stoessl, A. Jon. 2007. Gene Therapy for Parkinson's Disease: Early Data. *Lancet* 369:2056–58.

Stossel, Thomas P. 2005. Regulating Academic-Industrial Research Relationships: Solving Problems or Stifling Progress? *New England Journal of Medicine* 353 (10): 1060–65.

Swain, Sandra M. 2005. Aromatase Inhibitors: A Triumph of Translational Oncology. *New England Journal of Medicine* 353 (26): 2807–9.

Tanner, Lindsey (Associated Press). 2004. Group of Doctors Urges Aggressive Fight on Sepsis. *Boston Globe*. February 14.

Thomas, John R. 2004. *Scientific Research and the Experimental Use Privilege in Patent Law*. CRS Report for Congress. Congressional Research Service. October 28. http://ftp.fas.org/sgp/crs/RL32651.pdf (accessed June 22, 2007).

Thomas, John R., and Wendy H. Schacht. 2006. *Patent Reform: Innovation Issues*. CRS Report for Congress. Congressional Research Service. December 7. http://ipmall.info/hosted_resources/crs/RL32996-061207.pdf (accessed June 22, 2007).

———. 2007a. *Patent Reform: Innovation Issues*. CRS Report for Congress. January 17 (update). http://www.ipmall.info/hosted_resources/crs/RL32996-0701223.pdf (accessed July 26, 2007).

———. 2007b. *Patent Reform: Innovation Issues*. CRS Report for Congress. May 7 (update). http://fpc.state.gov/documents/organization/85624.pdf (accessed June 22, 2007).

Topol, Eric. 2004. Intensive Statin Therapy: A Sea Change in Cardiovascular Prevention. *New England Journal of Medicine* 350 (15): 1562–64.

Tuszynski, Mark H., Leon Thal, Mary Pay, David P. Salmon, Hoi Sang U, Roy Bakay, Piyush Patel, et al. 2005. A Phase 1 Clinical Trial of Nerve Growth Factor Gene Therapy for Alzheimer Disease. *Nature Medicine* 11 (5): 551–55.

U.S. Department of Health and Human Services. National Human Genome Research Institute. 2007. The ENCODE Project: ENCyclopedia of DNA Elements. http://www.genome.gov/10005107 (accessed July 30, 2007).

U.S. Department of Health and Human Services. National Institutes of Health. 2001. A Plan to Ensure Taxpayer Interests Are Protected. NIH Response to the Conference Report Request for a Plan to Ensure Taxpayers' Interests Are Protected.

July. http://www.ott.nih.gov/policy/policy_protect_text.html#b (accessed August 1, 2007).

———. 2004. Best Practices for the Licensing of Genomic Inventions. *Federal Register* 70 (68): 18413–15. http://www.ott.nih.gov/pdfs/70FR18413.pdf (accessed July 26, 2007).

———. 2007. Summary of FY 2077 President's Budget. http://officeofbudget.od.nih.gov/pdf/Press percent20info percent20final.pdf (accessed July 18, 2007).

U.S. Federal Trade Commission. 2002a. Business Perspectives on Patents: Biotech and Pharmaceuticals. Transcript of March 19, 2002, proceedings (hearings). http://www.ftc.gov/opp/intellect/020319trans.pdf (accessed June 19, 2007).

———. 2002b. Business Perspectives on Patents: Biotech and Pharmaceuticals. Transcript of February 26, 2002, proceedings (hearings). http://www.ftc.gov/opp/intellect/020226trans.pdf (accessed June 19, 2007).

———. 2003. To Promote Innovation: The Proper Balance of Competition and Patent Law and Policy. October. http://www.ftc.gov/os/2003/10/innovationrpt.pdf (accessed June 19, 2007).

U.S. House of Representatives. 2005. Judiciary Committee. Subcommittee on Courts, the Internet, and Intellectual Property. *Patent Act of 2005: Hearing before the Subcommittee on Courts, the Internet and Intellectual Property of the House Judiciary Committee, House of Representatives.* Testimony of Gary Griswold. 109th Cong., 1st sess. June 9. http://judiciary.house.gov/media/pdfs/printers/109th/21655.pdf (accessed July 27, 2007).

———. 2007a. Judiciary Committee. Subcommittee on Courts, the Internet, and Intellectual Property. *Hearing on H.R. 1908, The Patent Reform Act of 2007.* Statement of the Biotechnology Industry Organization. 110th Cong., 1st sess. April 26. http://patentsmatter.com/media/issue/resources/20070426_BIOStatement.pdf (accessed July 27, 2007).

———. 2007b. Judiciary Committee. Subcommittee on Courts, the Internet, and Intellectual Property. *Hearing on H.R. 1908, The Patent Reform Act of 2007.* Testimony of Gary Griswold. 110th Cong., 1st sess. April 26. http://judiciary.house.gov/media/pdfs/Griswold070426.pdf (accessed June 25, 2007).

———. 2007c. Small Business Committee. *Patent Reform: Impact on Small Venture-Backed Companies.* Testimony of John Neis. 110th Cong., 1st sess. March 29. http://www.nvca.org/pdf/House-SB-Patent-Testimony.pdf (accessed June 19, 2007).

U.S. Patent and Trademark Office (USPTO). 2001a. Utility Examination Guidelines. January 5. http://www.uspto.gov/web/offices/com/sol/notices/utilexmguide.pdf (accessed June 22, 2007).

———. 2001b. USPTO Publishes Final Guidelines for Determining Utility of Gene-Related Inventions. Press release. January 4. http://www.uspto.gov/web/offices/com/speeches/01-01.htm (accessed June 22, 2007).

———. 2002. *The 21st Century Strategic Plan.* Washington, D.C.: Government Printing Office.

———. 2006. *Performance and Accountability Report 2000–2006.* Washington, D.C.: Government Printing Office.

U.S. Senate. 2007a. Committee on the Judiciary. *Patent Reform: The Future of American Innovation.* Hearing before the Senate Committee on the Judiciary. Testimony of Kathryn L. Biberstein. 110th Cong., 1st sess. June 6. http://judiciary.senate.gov/testimony.cfm?id=2803&wit_id=6508 (accessed July 27, 2007).

———. 2007b. Committee on the Judiciary. *Patent Reform: The Future of American Innovation.* Hearing before the Senate Committee on the Judiciary. Testimony of Mary E. Doyle. 110th Cong., 1st sess. June 6. http://judiciary.senate.gov/testimony.cfm?id=2803&wit_id=6507 (accessed July 27, 2007).

Venter, J. Craig, Mark D. Adams, Eugene W. Myers, Peter W. Li, Richard J. Mural, Granger G. Sutton, Hamilton O. Smith, et al. 2001. The Sequence of the Human Genome. *Science* 1304 (291): 1304–52.

Vrazo, Fawn. 2005. Cancer Patients Staying at Work. *Philadelphia Inquirer.* April 11.

Wade, Nicholas. 2001. Swift Approval for a New Kind of Cancer Drug. *New York Times.* May 11.

Waldmeir, Patti. 2007. Supreme Court Declares War on Patents. *Financial Times.* May 2, 14.

Wall Street Journal. 2004. FDA Approves Avastin To Treat Colon Cancer. February 26.

Walsh, John P., Ashish Arora, and Wesley M. Cohen. 2003a. Effects of Research Tools Patents and Licensing on Biomedical Innovation. In *Patents in the Knowledge-Based Economy*, ed. Wesley M. Cohen and Stephen A. Merrill. Washington, D.C.: National Academies Press.

———. 2003b. Working Through the Patent Problem. *Science* 299 (5609): 1021.

Walsh, John P., Charlene Cho, and Wesley M. Cohen. 2005a. Patents, Material Transfers and Access to Research Inputs in Biomedical Research. Final Report to the National Academy of Sciences' Committee on Intellectual Property Rights in Genomic and Protein-Related Inventions. September 30. http://tigger.uic.edu/~jwalsh/WalshChoCohenFinal050922.pdf (accessed June 18, 2007).

———. 2005b. View from the Bench: Patents and Material Transfers. *Science* 309 (5743): 2002–3.

Weisfeldt, Myron L., and Susan J. Zieman. 2007. Advances in the Prevention and Treatment of Cardiovascular Disease. *Health Affairs* 26 (January–February): 25–37.

Wess, Ludger. 2005. Vaccines Getting Hot. *BioCentury.* October 17, A4–A9.

Yarris, Lynn, ed. 2004. *Biotechnology: Essays from Its Heartland.* BASIC (Bay Area Science and Innovation Consortium) and QB3 (California Institute for Quantitative Biomedical Research). June. http://www.qb3.org/pdfs/biotech1.pdf (accessed July 24, 2007).

Yeh, Brian T. 2007. An Overview of Recent U.S. Supreme Court Jurisprudence in Patent Law. CRS Report for Congress. Congressional Research Service. March 16. http://opencrs.cdt.org/rpts/RL33923_20070316.pdf (accessed June 22, 2007).

Zhou, Fangjun, Jeanne Santoli, Mark L. Messonnier, Hussain R. Yusuf, Abigail She-
   fer, Susan Y. Chu, Lance Rodewald, and Rafael Harpaz. 2005. Economic Evalua-
   tion of the 7-Vaccine Routine Childhood Immunization Schedule in the United
   States, 2001. *Archives of Pediatric and Adolescent Medicine* 159:1136–44.
Zimmerman, Rachel, and Scott Hensley. 2004. New Treatment Options For Breast
   Cancer: Promising Drug May Work Better Than Longtime Standard Tamoxifen.
   Wall Street Journal. March 11.
Zuniga, Cynthia A. 2007. *KSR v. Teleflex* in Brief. June 6. http://www.techlawforum.net/
   patent-reform/articles/ksr-v-teleflex-in-brief (accessed June 22, 2007).

About the Authors

Claude Barfield is a resident scholar at the American Enterprise Institute. His most recent publications are *High-Tech Protectionism: The Irrationality of Anti-Dumping Laws* and *Free Trade, Sovereignty, Democracy: The Future of the World Trade Organization.* Earlier studies in the area of intellectual property include *Parallel Trade in the Pharmaceutical Industry: Implications for Innovation, Consumer Welfare and Health Policy* (with Mark Groombridge), and *The Economic Case for Copyright Owner Control over Parallel Imports* (with Mark Groombridge).

John E. Calfee has been a resident scholar at the American Enterprise Institute since 1995. After receiving his Ph.D. in economics from the University of California at Berkeley, he spent several years in the Bureau of Economics at the Federal Trade Commission and taught at the business schools of the University of Maryland and Boston University. Much of his recent work has focused on the pharmaceutical industry, with special attention to drug development, FDA regulation, and the economic underpinnings of both biotech and traditional drug R&D. His work on pharmaceuticals has appeared in scholarly journals such as the *Annals of Internal Medicine*, the *Journal of Law and Economics*, *Health Affairs*, and *Pharmacoeconomics*, as well as in the *Wall Street Journal*, *The American*, and other popular press publications.